T0270913

LONDON MATHEMATICAL SOCIETY STUDENT TEXTS

Managing Editor: Ian J. Leary,
Mathematical Sciences, University of Southampton, UK

56 Logic, induction and sets, THOMAS FORSTER
57 Introduction to Banach algebras, operators and harmonic analysis, GARTH DALES *et al.*
58 Computational algebraic geometry, HAL SCHENCK
59 Frobenius algebras and 2-D topological quantum field theories, JOACHIM KOCK
60 Linear operators and linear systems, JONATHAN R. PARTINGTON
61 An introduction to noncommutative Noetherian rings: Second edition, K. R. GOODEARL & R. B. WARFIELD, JR
62 Topics from one-dimensional dynamics, KAREN M. BRUCKS & HENK BRUIN
63 Singular points of plane curves, C. T. C. WALL
64 A short course on Banach space theory, N. L. CAROTHERS
65 Elements of the representation theory of associative algebras I, IBRAHIM ASSEM, DANIEL SIMSON & ANDRZEJ SKOWROŃSKI
66 An introduction to sieve methods and their applications, ALINA CARMEN COJOCARU & M. RAM MURTY
67 Elliptic functions, J. V. ARMITAGE & W. F. EBERLEIN
68 Hyperbolic geometry from a local viewpoint, LINDA KEEN & NIKOLA LAKIC
69 Lectures on Kähler geometry, ANDREI MOROIANU
70 Dependence logic, JOUKU VÄÄNÄNEN
71 Elements of the representation theory of associative algebras II, DANIEL SIMSON & ANDRZEJ SKOWROŃSKI
72 Elements of the representation theory of associative algebras III, DANIEL SIMSON & ANDRZEJ SKOWROŃSKI
73 Groups, graphs and trees, JOHN MEIER
74 Representation theorems in Hardy spaces, JAVAD MASHREGHI
75 An introduction to the theory of graph spectra, DRAGOŠ CVETKOVIĆ, PETER ROWLINSON & SLOBODAN SIMIĆ
76 Number theory in the spirit of Liouville, KENNETH S. WILLIAMS
77 Lectures on profinite topics in group theory, BENJAMIN KLOPSCH, NIKOLAY NIKOLOV & CHRISTOPHER VOLL
78 Clifford algebras: An introduction, D. J. H. GARLING
79 Introduction to compact Riemann surfaces and dessins d'enfants, ERNESTO GIRONDO & GABINO GONZÁLEZ-DIEZ
80 The Riemann hypothesis for function fields, MACHIEL VAN FRANKENHUIJSEN
81 Number theory, Fourier analysis and geometric discrepancy, GIANCARLO TRAVAGLINI
82 Finite geometry and combinatorial applications, SIMEON BALL
83 The geometry of celestial mechanics, HANSJÖRG GEIGES
84 Random graphs, geometry and asymptotic structure, MICHAEL KRIVELEVICH *et al.*
85 Fourier analysis: Part I – Theory, ADRIAN CONSTANTIN
86 Dispersive partial differential equations, M. BURAK ERDOĞAN & NIKOLAOS TZIRAKIS
87 Riemann surfaces and algebraic curves, R. CAVALIERI & E. MILES
88 Groups, languages and automata, DEREK F. HOLT, SARAH REES & CLAAS E. RÖVER
89 Analysis on Polish spaces and an introduction to optimal transportation, D. J. H. GARLING
90 The homotopy theory of (∞,1)-categories, JULIA E. BERGNER
91 The block theory of finite group algebras I, M. LINCKELMANN
92 The block theory of finite group algebras II, M. LINCKELMANN

London Mathematical Society Student Texts 93

Semigroups of Linear Operators

With Applications to Analysis, Probability and Physics

DAVID APPLEBAUM

University of Sheffield

CAMBRIDGE
UNIVERSITY PRESS

CAMBRIDGE
UNIVERSITY PRESS

University Printing House, Cambridge CB2 8BS, United Kingdom

One Liberty Plaza, 20th Floor, New York, NY 10006, USA

477 Williamstown Road, Port Melbourne, VIC 3207, Australia

314-321, 3rd Floor, Plot 3, Splendor Forum, Jasola District Centre, New Delhi - 110025, India

79 Anson Road, #06-04/06, Singapore 079906

Cambridge University Press is part of the University of Cambridge.

It furthers the University's mission by disseminating knowledge in the pursuit of education, learning and research at the highest international levels of excellence.

www.cambridge.org
Information on this title: www.cambridge.org/9781108483094
DOI: 10.1017/9781108672641

© David Applebaum 2019

First published 2019

A catalogue record for this publication is available from the British Library

Library of Congress Cataloging in Publication data
Names: Applebaum, David, 1956– author.
Title: Semigroups of linear operators : with applications to analysis, probability and physics / David Applebaum (University of Sheffield).
Description: Cambridge ; New York, NY : Cambridge University Press, 2019. I Series: London Mathematical Society student texts ; 93 I Includes bibliographical references and index.
Identifiers: LCCN 2019010021 I ISBN 9781108483094 (alk. paper)
Subjects: LCSH: Linear operators – Textbooks. I Operator theory – Textbooks. I Semigroups – Textbooks. I Group theory – Textbooks.
Classification: LCC QA329.2 .A687 2019 I DDC 515/.7246–dc23
LC record available at https://lccn.loc.gov/2019010021

ISBN 978-1-108-48309-4 Hardback
ISBN 978-1-108-71637-6 Paperback

Additional resources for this publication at www.cambridge.org/9781108483094.

This book is dedicated to the mathematical community
of Costa Rica. Long may it flourish.

"I hail a semigroup when I see one, and I seem to see them everywhere!"

(Einar Hille, Foreword to [45])

Contents

Introduction   *page* 1

1 Semigroups and Generators   9
   1.1 Motivation from Partial Differential Equations   9
   1.2 Definition of a Semigroup and Examples   10
   1.3 Unbounded Operators and Generators   15
       1.3.1 Unbounded Operators and Density of Generators   15
       1.3.2 Differential Equations in Banach Space   18
       1.3.3 Generators as Closed Operators   20
       1.3.4 Closures and Cores   21
   1.4 Norm-Continuous Semigroups   23
   1.5 The Resolvent of a Semigroup   25
       1.5.1 The Resolvent of a Closed Operator   25
       1.5.2 Properties of the Resolvent of a Semigroup   27
   1.6 Exercises for Chapter 1   29

2 The Generation of Semigroups   31
   2.1 Yosida Approximants   31
   2.2 Classifying Generators   33
   2.3 Applications to Parabolic PDEs   38
       2.3.1 Bilinear Forms, Weak Solutions and the
             Lax–Milgram Theorem   39
       2.3.2 Energy Estimates and Weak Solutions to the Elliptic
             Problem   40
       2.3.3 Semigroup Solution of the Parabolic Problem   41
   2.4 Exercises for Chapter 2   43

3 Convolution Semigroups of Measures   46
   3.1 Heat Kernels, Poisson Kernels, Processes and Fourier
       Transforms   46

3.1.1   The Gauss–Weierstrass Function and the Heat Equation   46
3.1.2   Brownian Motion and Itô's Formula   50
3.1.3   The Cauchy Distribution, the Poisson Kernel and
         Laplace's Equation   52
3.2   Convolution of Measures and Weak Convergence   55
3.2.1   Convolution of Measures   55
3.2.2   Weak Convergence   57
3.3   Convolution Semigroups of Probability Measures   59
3.4   The Lévy–Khintchine Formula   64
3.4.1   Stable Semigroups   69
3.4.2   Lévy Processes   71
3.5   Generators of Convolution Semigroups   73
3.5.1   Lévy Generators as Pseudo-Differential Operators   76
3.6   Extension to L^p 78
3.7   Exercises for Chapter 3   81

4   Self-Adjoint Semigroups and Unitary Groups   83
4.1   Adjoint Semigroups and Self-Adjointness   83
4.1.1   Positive Self-Adjoint Operators   85
4.1.2   Adjoints of Semigroups on Banach Spaces   88
4.2   Self-Adjointness and Convolution Semigroups   89
4.3   Unitary Groups, Stone's Theorem   91
4.4   Quantum Dynamical Semigroups   97
4.5   Exercises for Chapter 4   100

5   Compact and Trace Class Semigroups   102
5.1   Compact Semigroups   102
5.2   Trace Class Semigroups   104
5.2.1   Hilbert–Schmidt and Trace Class Operators   104
5.2.2   Trace Class Semigroups   108
5.2.3   Convolution Semigroups on the Circle   109
5.2.4   Quantum Theory Revisited   112
5.3   Exercises for Chapter 5   114

6   Perturbation Theory   116
6.1   Relatively Bounded and Bounded Perturbations   116
6.1.1   Contraction Semigroups   116
6.1.2   Analytic Semigroups   119
6.2   The Lie–Kato–Trotter Product Formula   120
6.3   The Feynman–Kac Formula   123

	6.3.1	The Feynman–Kac Formula via the Lie–Kato–Trotter Product Formula	123
	6.3.2	The Feynman–Kac Formula via Itô's Formula	125
6.4	Exercises for Chapter 6		127

7 Markov and Feller Semigroups — 129
7.1 Definitions of Markov and Feller Semigroups — 129
7.2 The Positive Maximum Principle — 133
 7.2.1 The Positive Maximum Principle and the Hille–Yosida–Ray Theorem — 133
 7.2.2 Crash Course on Distributions — 135
 7.2.3 The Courrège Theorem — 136
7.3 The Martingale Problem — 144
 7.3.1 Sub-Feller Semigroups — 147

8 Semigroups and Dynamics — 149
8.1 Invariant Measures and Entropy — 149
 8.1.1 Invariant Measures — 149
 8.1.2 Entropy — 151
8.2 Semidynamical Systems — 152
 8.2.1 Koopmanism — 152
 8.2.2 Dynamical Systems and Differential Equations — 157
8.3 Approaches to Irreversibility — 161
 8.3.1 Dilations of Semigroups — 162
 8.3.2 The Misra–Prigogine–Courbage Approach — 164

9 Varopoulos Semigroups — 166
9.1 Prelude – Fractional Calculus — 166
9.2 The Hardy–Littlewood–Sobolev Inequality and Riesz Potential Operators — 168
9.3 Varopoulos Semigroups — 169
9.4 Varopoulos's Theorem — 173
9.5 Nash Inequality, Symmetric Diffusion Semigroups and Heat Kernels — 175

Notes and Further Reading — 179

Appendix A **The Space $C_0(\mathbb{R}^d)$** — 182

Appendix B **The Fourier Transform** — 186

Appendix C **Sobolev Spaces** — 190

Appendix D   **Probability Measures and Kolmogorov's Theorem on Construction of Stochastic Processes**   194

Appendix E   **Absolute Continuity, Conditional Expectation and Martingales**   197
   E.1   The Radon–Nikodym Theorem   197
   E.2   Conditional Expectation   200
   E.3   Martingales   203

Appendix F   **Stochastic Integration and Itô's Formula**   205
   F.1   Stochastic Integrals   205
   F.2   Itô's Formula   208

Appendix G   **Measures on Locally Compact Spaces – Some Brief Remarks**   212

   References   214
   Index   219

Introduction

In May 2017, I was privileged to be invited to give a mini-course of lectures at the University of Costa Rica in San José. The participants were mainly a group of advanced undergraduate students who had already taken courses in functional analysis and measure theory. After some discussions with my hosts, we decided that I would lecture on the topic of this book, which has since evolved from the manuscript that I prepared for the lectures into the volume that you are now reading.

Why semigroups? First I should emphasise that I am not an "expert" in this subject, but despite this, it seems that I have, in some sense, spent my entire research career working on this topic, in that whichever problem I happen to be working on, there is always a semigroup either blatantly hogging the limelight in the foreground, or "lurking in the background".[1] The subject is undeniably very attractive from the point of view of intellectual beauty. It is a delightful tour-de-force of fascinating mathematical ideas, which forms a very natural second course in functional analysis, for those who have already had some grounding in the structure of Banach and Hilbert spaces, and associated linear operators. But it is also very deeply connected with applications that both illustrate the theory itself, and also provide the impetus for new theoretical developments. These include, but are not restricted to, partial differential equations, stochastic processes, dynamical systems and quantum theory, and each of these topics features within the current book.

There are already many excellent existing books on semigroup theory and applications, so why publish another? Firstly, I have tried to write a book that exhibits the spirit of my lectures. So I am assuming that the primary readership will consist of final-year undergraduates, MSc students and beginning

---

[1] Apologies to Nick Bingham for stealing one of his favourite catchphrases.

PhD students, and there is a corresponding pedagogic approach that assumes somewhat less sophistication and experience on behalf of the reader than is usually found in books on this subject. Secondly, I have tried to bring interesting classes of examples into play at an early stage; and this means that there is far more interplay between functional analysis and other areas of analysis – in particular, measure theoretic probability – than in other accounts. Of course, much interdisciplinary research involves the interaction of one or more branches of mathematics, and I hope that readers will benefit by being exposed to this way of thinking within the current text.

The semigroups that we are concerned with will be families of bounded linear operators $(T_t, t \geq 0)$ acting in a real or complex Banach space E, which satisfy the semigroup property $T_{s+t} = T_s T_t$ for all $t \geq 0$, with T_0 being the identity operator on E, and which have nice continuity properties. We will see that there is then a linear operator A acting in E, which is typically not bounded, which is obtained by differentiating the semigroup at $t = 0$. We call A the generator of the semigroup, and one of the key themes of the first two chapters is the interplay between the semigroup and the generator. We do not assume prior knowledge of the theory of unbounded operators, and seek to develop what is needed as we go along. It is natural to interpret the parameter t as describing the flow of time, and the semigroup itself as the dynamical time-evolution of some system of interest. Then the semigroup and its generator represent the global and local descriptions of the dynamics (respectively). Many examples of this type come from partial differential equations (PDEs), where the generator A is typically a second-order (elliptic) differential operator. Other important examples come from the world of stochastic processes, where the semigroup is obtained by averaging over all possible trajectories of some random dynamical evolution. A fascinating aspect of working in this area is the appreciation that these application areas are not distinct, so we can and will approach the same phenomenon from the point of view of semigroup theory, probability theory and PDEs.

The first chapter and first two sections of the second are designed to give a pretty rigorous and thorough introduction to the basic concepts of the theory. In particular, the first part of Chapter 2 presents proofs of the three key theorems that give necessary and sufficient conditions for a linear operator A to be the generator of a semigroup, these being the Feller–Miyadera–Phillips, Hille–Yosida and Lumer–Phillips theorems. Having reached this point, we do not feel overly constrained to give fully mathematically rigorous accounts of what follows. Our goal is more to present a wide range of different topics so that readers can get an introduction to the landscape, but to give only partial proofs where there are too many technical details, or even just heuristic arguments.

In the latter cases, we are of course very careful to give precise references to where detailed proofs can be found.

Continuing our brief tour of the content of the book, the second part of Chapter 2 deals with partial differential equations, and the main point is to show that solutions of second-order parabolic equations can be represented as semigroup actions. Chapter 3 is about semigroups of operators that are obtained from convolution semigroups of measures. To some extent, it is a companion piece to Chapters 1 and 3 of my earlier book [6], but in that work, the emphasis was on the underlying stochastic processes, whereas here we take a more analytic perspective and place the semigroup in centre ground. We present a proof (at least in outline) of the famous Lévy–Khintchine formula which characterises the convolution semigroup through its Fourier transform. The generators are then conveniently represented as pseudo-differential operators, and we also include an introduction to this important theme of modern analysis for those readers who have not met them before.

Three of the most important classes of operators in Hilbert space that are commonly encountered are the self-adjoint, compact and trace class. In Chapter 4, we study self-adjoint semigroups, but also unitary groups, that are generated by iA, where A is self-adjoint. This two-way relationship between skew-adjoint generators and unitary groups, is the content of Stone's theorem, which is proved in this chapter. This paves the way for a discussion of quantum mechanics, as the key Schrödinger equation is an infinitesimal expression of Stone's theorem. Being group dynamics, this is reversible; but we also discuss irreversible dynamics in the quantum context. We do not prove the celebrated Gorini–Kossakowski–Sudarshan–Lindlad theorem that classifies generators in this context, but we do give some probabilistic insight into how such operators can arise. Chapter 5 is concerned with compact and trace class semigroups. We investigate eigenfunction expansions for the semigroup and its kernel (when it acts as an integral operator), and also meet the important Mercer's theorem that relates the trace to the kernel. In both Chapters 4 and 5, the convolution semigroups studied in Chapter 3 are put to work to yield important classes of examples.

In Chapter 6, we take a brief look at perturbation theory, and conclude by giving two derivations of the celebrated Feynman–Kac formula, one based on the Lie–Kato–Trotter product formula, which is proved herein, and the other using Itô calculus. Chapter 7 returns to the theme of Chapter 3, but in greater generality as the context is now Markov and Feller semigroups. Here we give a partial proof of the Hille–Yosida–Ray theorem that gives necessary and sufficient conditions for an operator to generate a positivity-preserving contraction semigroup. One of these conditions is the positive maximum theorem, and we include a full proof of the Courrège theorem that gives the characteristic form of operators that obey this principle. Our proof uses some ideas from the theory of distributions, and again we give a brief self-contained synopsis of all the material that we'll need.

In Chapter 8, we look more carefully at the relationship between semigroups and (semi)-dynamical systems. The main idea here is that we have a group (say) of transformations on a locally compact space S that expresses some dynamical law. We pull back these transformations to a group of operators acting in a suitable L^2-space. The aim is then to study the original dynamical system from the point of view of the group of operators. This chapter also contains a brief discussion of mathematical models of the origins of irreversibility. Here we try to engage, in a mathematical way, with the fascinating question: is irreversible evolution a fundamental aspect of the way in which Nature works, or just a feature of our inability to see into the real nature of the dynamics, which is reversible?

Finally in Chapter 9, we introduce a class of semigroups on function spaces, called Varopoulos semigroups, after N. Varopoulos who invented them. They are closely related to ultracontractive semigroups. We prove that the Riesz potential operators that occur as the Mellin transforms of Varopoulos semigroups satisfy the Hardy–Littlewood–Sobolev inequality, and obtain some Sobolev inequalities as a special case. We then show that second-order elliptic partial differential operators on bounded regions generate Varopoulos semigroups, and prove the famous Nash inequality along the way.

As can be seen from this synopsis, as well as introducing and developing semigroup theory, our journey enables us to touch on a number of important topics within contemporary analysis. However, in order to keep this introductory book to manageable size, many interesting topics were omitted, such as analytic semigroups,[2] subordination and Dirichlet forms. Of course readers wanting to know more about these and other topics will find ample resources in the bibliography. Where there is more than one edition of a book listed there, I have generally used the more recent one. Each of the first six chapters of the book concludes with a set of exercises. Solutions to these will be made available at www.cambridge.org/9781108483094. The proofs of some theorems that are either only stated in the main text (such as the Lax–Milgram theorem in Chapter 2), or which lie outside the central scope of the book, but are nonetheless both interesting and important (such as the basic criteria for self-adjointness in Chapter 4), appear in the exercises with sufficient guidance for readers to be able to construct

---

[2] These are in fact given a very short treatment in subsection 6.1.2.

these for themselves. When in the main text, you encounter a reference to Problem x.y, this means the yth problem in the exercises at the end of Chapter x.

It is a pleasure to thank my former PhD student (now a lecturer at the University of Costa Rica) Christian Fonseca Mora, for inviting me there and for working so hard to make my visit enjoyable. I will never forget the warm and generous hospitality extended to me by Christian, his wife Yeime and their extended family. It was also heartening to meet so many keen students, who were exceptionally well-prepared for my lectures, and so hungry to learn modern analysis. I hope that I was able to satisfy their appetites, at least for a short while. Thanks are also due to my PhD students Rosemary Shewell Brockway and Trang Le Ngan, and my colleagues Nic Freeman and Koji Ohkitani, who attended a short informal course based on this material at the University of Sheffield in spring 2018. Last, but not least (on the academic side), I would like to thank both Christian Fonseca Mora and Gergely Bodó (a former undergraduate student at the University of Sheffield) for their careful reading of the manuscript, enabling me to correct many typos and minor errors. Finally it is a pleasure to thank all the hard-working staff at Cambridge University Press who have helped to transform my manuscript into the high-quality book (or e-book) that you are currently reading, particularly the editors Roger Astley and Clare Dennison, project manager Puviarassy Kalieperumal, and copyeditor Bret Workman.

Guide to Notation and a Few Useful Facts

If S is a set, S^c denotes its complement. If T is another set, then $S \backslash T := S \cap T^c$. If A is a finite set, then the number of elements in A is $\#A$. If A is a non-empty subset of a topological space, then $\overline{A}$ is its closure. If S is a metric space, with metric d, then for each $x \in S$, $B_r(x) := \{y \in S; d(x, y) < r\}$ is the open ball of radius $r > 0$, centred at x.

$\mathbb{N}, \mathbb{Z}, \mathbb{Q}, \mathbb{R}, \mathbb{C}$ are the sets of natural numbers, integers, rational numbers, real numbers and complex numbers (respectively). $\mathbb{Z}_+ := \mathbb{N} \cup \{0\}$. In this short section we use F to denote $\mathbb{R}$ or $\mathbb{C}$.

If S is a topological space, then the *Borel* σ-algebra of S will be denoted $\mathcal{B}(S)$. It is the smallest σ-algebra of subsets of S which contains all the open sets. Sets in $\mathcal{B}(S)$ are called *Borel sets*. $\mathbb{R}$ and $\mathbb{C}$ will always be assumed to be equipped with their Borel σ-algebras, and measurable functions from $(S, \mathcal{B}(S))$ to $(F, \mathcal{B}(F))$ are sometimes called *Borel measurable*. Similarly, a measure defined on $(S, \mathcal{B}(S))$ is called a *Borel measure*.

If S is a locally compact Hausdorff space[3] (we will not meet these general spaces until Chapter 6), then $B_b(S, F)$ is the linear space (with the usual pointwise operations of addition and scalar multiplication) of all bounded Borel measurable functions from S to F. It is an F-Banach space under the supremum norm $||f||_\infty := \sup_{x \in S} |f(x)|$ for $f \in B_b(S, F)$. The space of bounded continuous functions from S to F is denoted $C_b(S, F)$. It is a closed linear subspace of $B_b(S, F)$, and so an F-Banach space in its own right. A function f from S to F is said to *vanish at infinity* if given any $\epsilon > 0$ there exists a compact set K in S so that $|f(x)| < \epsilon$ whenever $x \in K^c$. The space $C_0(S, F)$ of all continuous F-valued functions on S which vanish at infinity[4] is a closed linear subspace of $B_b(S, F)$ (and of $C_b(S, F)$), and so is also an F-Banach space in its own right. The *support* of an F-valued function f defined on S is the closure of the set $\{x \in S; f(x) \neq 0\}$ and it is denoted $\mathrm{supp}(f)$. The linear space $C_c(S, F)$ of all continuous F-valued functions on S with compact support is a dense subspace of $C_0(S, F)$. When $F = \mathbb{R}$, we usually write $B_b(S) := B_b(S, \mathbb{R})$, $C_0(S) := C_0(S, \mathbb{R})$ etc., and this will be our default assumption.

Throughout this book "smooth" means "infinitely differentiable". We write $C(\mathbb{R}^d)$ for the linear space of all continuous real-valued functions on $\mathbb{R}^d$. If $n \in \mathbb{N}, C^n(\mathbb{R}^d)$ is the linear space of all n-times real-valued differentiable functions on $\mathbb{R}^d$ that have continuous partial derivatives to all orders. We define $C^\infty(\mathbb{R}^d) := \bigcap_{n \in \mathbb{N}} C^n(\mathbb{R}^d)$, and $C_c^n(\mathbb{R}^d) := C^n(\mathbb{R}^d) \cap C_c(\mathbb{R}^d)$ for all $n \in \mathbb{N} \cup \{\infty\}$.

If $n \in \mathbb{N}$, then $M_n(F)$ is the F-algebra of all $n \times n$ matrices with values in F. The identity matrix in $M_n(F)$ is denoted by I_n. If $A \in M_n(F)$, then A^T denotes its transpose and $A^* = \overline{A^T}$ is its adjoint. The trace of a square matrix A, i.e., the sum of its diagonal entries, is denoted by $\mathrm{tr}(A)$, and its determinant is $\det(A)$. The matrix A is said to be *non-negative definite* if $x^T A x \geq 0$ for all $x \in \mathbb{R}^n$, and *positive definite* if $x^T A x > 0$ for all $x \in \mathbb{R}^n \setminus \{0\}$.

If (S, Σ, μ) is a measure space and $f : S \to F$ is an integrable function, we often write the Lebesgue integral $\int_S f(x)\mu(dx)$ as $\mu(f)$. For $1 \leq p < \infty, L^p(S) := L^p(S, \mathbb{C}) = L^p(S, \Sigma, \mu; F; \mathbb{C})$ is the usual L^p space of equivalence classes of complex-valued functions that agree almost everywhere with respect to μ for which

$$||f||_p = \left(\int_S |f(x)|^p \mu(dx) \right)^{\frac{1}{p}} < \infty$$

---

[3] Readers who have not yet learned these topological notions are encouraged to take $S \subseteq \mathbb{R}^d$.
[4] This space will play an important role in this book. Some of its key properties are proved in Appendix A.

for all $f \in L^p(S)$. $L^p(S)$ is a Banach space with respect to the norm $|| \cdot ||_p$, and $L^2(S)$ is a Hilbert space with respect to the inner product

$$\langle f, g \rangle := \int_S f(x)\overline{g(x)}\mu(dx)$$

for $f, g \in L^2(S)$. The spaces $L^p(S, \Sigma, \mu; F; \mathbb{R})$ are defined similarly. For applications to probability theory and PDEs, we will usually write $L^p(S) := L^p(S, \mathbb{R})$, but if we deal with Fourier transforms or (in the case $p = 2$) quantum mechanics, then we will need complex-valued functions.

The *indicator function* 1_A of $A \in \Sigma$ is defined as follows:

$$1_A(x) = \begin{cases} 1 & \text{if } x \in A \\ 0 & \text{if } x \notin A. \end{cases}$$

If μ is σ-finite, and ν is a finite measure on (S, Σ), we write $\nu \ll \mu$ if ν is absolutely continuous with respect to μ, and $\dfrac{d\nu}{d\mu}$ is the corresponding Radon–Nikodym derivative (see Appendix E).

If $(\Omega, \mathcal{F}, P)$ is a probability space and $X : \Omega \rightarrow \mathbb{R}$ is a random variable (i.e., a measurable function from $(\Omega, \mathcal{F})$ to $(\mathbb{R}, \mathcal{B}(\mathbb{R}))$) that is also integrable in that $\int_\Omega |X(\omega)| P(d\omega) < \infty$, then its *expectation* is defined to be

$$\mathbb{E}(X) := \int_\Omega X(\omega) P(d\omega).$$

If $T : V_1 \rightarrow V_2$ is a linear mapping between F-vector spaces V_1 and V_2, then $\text{Ker}(T)$ is its kernel and $\text{Ran}(T)$ is its range. If $T : H_1 \rightarrow H_2$ is a bounded linear operator between F-Hilbert spaces H_i having inner products $\langle \cdot, \cdot \rangle_i$ ($i = 1, 2$), its *adjoint* is the unique bounded linear operator $T^* : H_2 \rightarrow H_1$ for which

$$\langle T^*\psi, \phi \rangle_1 = \langle \psi, T\phi \rangle_2,$$

for all $\phi \in H_1, \psi \in H_2$. The bounded linear operator $U : H_1 \rightarrow H_2$ is said to be *unitary* if it is both an isometry and a co-isometry (i.e., U^* is also an isometry). Equivalently it is an isometric isomorphism for which $U^{-1} = U^*$.

If E is an F-Banach space, then $\mathcal{L}(E)$ will denote the algebra of all bounded linear operators on E. $\mathcal{L}(E)$ is a Banach space with respect to the operator norm

$$||T|| = \sup\{||Tx||; x \in H, ||x|| = 1\}.$$

Note that the algebra $M_n(F)$ may be realised as $\mathcal{L}(F^n)$. The (topological) dual space E' of E is the linear space of all bounded linear maps (often called linear functionals) from E to F. It is a Banach space with respect to the norm:

$$||l|| = \sup\{|l(x)|; x \in E, ||x|| = 1\}.$$

We typically use $\langle \cdot, \cdot \rangle$ to indicate the dual pairing between E' and E, so if $l \in E', x \in E$,

$$l(x) = \langle x, l \rangle.$$

If H is a Hilbert space and $x, y \in H$ are orthogonal so that $\langle x, y \rangle = 0$, we sometimes write $x \perp y$. In the main part of the book, we will always write $\langle \cdot, \cdot \rangle_{\mathbb{C}}$ as $\langle \cdot, \cdot \rangle$. It is perhaps worth emphasising that all inner products on complex vector spaces are linear on the left and conjugate-linear on the right, which is standard in mathematics (but not in physics). We use $\mathcal{H}$ instead of H to denote our Hilbert space whenever there is an operator called H (typically a quantum mechanical Hamiltonian) playing a role within that section of the text.

The inner product (scalar product) of $x, y \in \mathbb{R}^d$ is always written $x \cdot y$, and the associated norm is $|x| := \left(\sum_{i=1}^d x_i^2 \right)^{\frac{1}{2}}$, for $x = (x_1, \ldots, x_d)$.

If $a, b \in \mathbb{R}$, then $a \wedge b := \min\{a, b\}$.

Throughout the book, we will use standard notation for partial differential operators. Let $\alpha = (\alpha_1, \ldots, \alpha_d)$ be a *multi-index*, so that $\alpha \in (\mathbb{N} \cup \{0\})^d$. We define $|\alpha| = \alpha_1 + \cdots + \alpha_d$ and

$$D^\alpha = \frac{1}{i^{|\alpha|}} \frac{\partial^{\alpha_1}}{\partial x_1^{\alpha_1}} \cdots \frac{\partial^{\alpha_d}}{\partial x_d^{\alpha_d}}.$$

Similarly, if $x = (x_1, \ldots, x_d) \in \mathbb{R}^d$, then $x^\alpha = x_1^{\alpha_1} \cdots x_d^{\alpha_d}$. For ease of notation, we will usually write ∂_i instead of $\frac{\partial}{\partial x_i}$ for $i = 1, \ldots, d$.

If S is a set, then we use ι for the identity mapping, $\iota(x) = x$, for all $x \in S$.

If f is a real or complex-valued function on $\mathbb{R}^d$ and $a \in \mathbb{R}^d$, $\tau_a f$ is the shifted function, defined by

$$(\tau_a f)(x) = f(x + a),$$

for all $x \in \mathbb{R}^d$.

1

Semigroups and Generators

1.1 Motivation from Partial Differential Equations

Consider the following initial value problem on $\mathbb{R}^d$:

$$\frac{\partial u(t, x)}{\partial t} = \sum_{i,j=1}^{d} a_{ij}(x)\partial_{ij}^2 u(t, x) + \sum_{i=1}^{d} b_i(x)\partial_i u(t, x) - c(x)u(t, x),$$

$$u(0, \cdot) = f(\cdot), \tag{1.1.1}$$

where $f \in C_c^2(\mathbb{R}^d)$. Here we have used the simplifying notation ∂_i for $\frac{\partial}{\partial x_i}$ and ∂_{ij} for $\frac{\partial^2}{\partial x_i \partial x_j}$. We will assume that c, b_i and $a_{j,k}$ are bounded smooth (i.e., infinitely differentiable) functions on $\mathbb{R}^d$ for each $i, j, k = 1, \ldots, d$ with $c \geq 0$, and that the matrix-valued function $a = (a_{ij})$ is uniformly elliptic in that there exists $K > 0$ so that

$$\inf_{x \in \mathbb{R}^d} a(x)\xi \cdot \xi \geq K|\xi|^2 \text{ for all } \xi \in \mathbb{R}^d,$$

with $(a_{ij}(x))$ being a symmetric matrix for each $x \in \mathbb{R}^d$.

The point of these conditions is to ensure that (1.1.1) has a unique solution.[1] We make no claims that they are, in any sense, optimal. We rewrite (1.1.1) as an abstract ordinary differential equation:

$$\frac{du_t}{dt} = Au_t,$$

$$u_0 = f, \tag{1.1.2}$$

---

[1] See, e.g., Chapter 6, section 5 of Engel and Nagel [31].

where A is the linear operator

$$(Ag)(x) = \sum_{i,j=1}^{d} a_{ij}(x)\partial^2_{ij}g(x) + \sum_{i=1}^{d} b_i(x)\partial_i g(x) - c(x)g(x),$$

acting on the linear space of all functions $g \colon \mathbb{R}^d \to \mathbb{R}$ that are at least twice differentiable.

The original PDE (1.1.1) acted on functions of both time and space and the solution u is a function from $[0, \infty) \times \mathbb{R}^d \to \mathbb{R}$. In (1.1.2), we have hidden the spatial dependence within the structure of the operator A and our solution is now a family of functions $u_t \colon E \to E$, where E is a suitable space of functions defined on $\mathbb{R}^d$. It is tempting to integrate (1.1.2) naively, write the solution as

$$u_t = e^{tA}f, \qquad (1.1.3)$$

and seek to interpret e^{tA} as a linear operator in E. From the discussion above, it seems that E should be a space of twice-differentiable functions, but such spaces do not have a rich structure from a functional analytic perspective. Our goal will be to try to make sense of e^{tA} when E is a Banach space, such as $C_0(\mathbb{R}^d)$, or $L^p(\mathbb{R}^d)$ (for $p \geq 1$). If we are able to do this, then writing $T_t := e^{tA}$, we should surely have that for all $s, t \geq 0$,

$$T_{s+t} = T_s T_t \text{ and } T_0 = I.$$

We would also expect to be able to recapture A from the mapping $t \to T_t$ by $A = \frac{d}{dt}\big|_{t=0} T_t$. Note that if $E = L^2(\mathbb{R}^d)$, then we are dealing with operators on a Hilbert space, and if we impose conditions on the b_i's and a_{jk}'s such that A is self-adjoint,[2] then we should be able to use spectral theory to write

$$A = \int_{\sigma(A)} \lambda \, dE(\lambda), \; T_t = \int_{\sigma(A)} e^{t\lambda} dE(\lambda),$$

where $\sigma(A)$ is the spectrum of A. We now seek to turn these musings into a rigorous mathematical theory.

1.2 Definition of a Semigroup and Examples

Most of the material given below is standard. There are many good books on semigroup theory and we have followed Davies [27] very closely.

---

[2] This is non-trivial as A is not a bounded operator, see below.

Let E be a real or complex Banach space and $\mathcal{L}(E)$ be the algebra of all bounded linear operators on E. A C_0-semigroup[3] on E is a family of bounded, linear operators $(T_t, t \geq 0)$ on E for which

(S1) $T_{s+t} = T_s T_t$ for all $s, t \geq 0$,
(S2) $T_0 = I$,
(S3) the mapping $t \to T_t \psi$ from $[0, \infty)$ to E is continuous for all $\psi \in E$.

We briefly comment on these. (S1) and (S2) are exactly what we expect when we try to solve differential equations in Banach space – we saw this in section 1.1. We need (S3) since without it, as we will see, it is very difficult to do any analysis. For many classes of examples that we consider, $(T_t, t \geq 0)$ will be a C_0-semigroup such that T_t is a contraction[4] for all $t > 0$. In this case, we say that $(T_t, t \geq 0)$ is a *contraction semigroup*. Finally, if only (S1) and (S2) (but not (S3)) are satisfied we will call $(T_t, t \geq 0)$ an *algebraic operator semigroup*, or *AO semigroup*, for short.

The condition (S3) can be simply expressed as telling us that the mapping $t \to T_t$ is strongly continuous in $\mathcal{L}(E)$. We now show that we can replace it with the seemingly weaker condition:

(S3)′ The mapping $t \to T_t \psi$ from $\mathbb{R}^+$ to E is continuous at $t = 0$ for all $\psi \in E$.

Proposition 1.2.1 *If $(T_t, t \geq 0)$ is a family of bounded linear operators on E satisfying (S1) and (S2), then it satisfies (S3) if and only if it satisfies (S3)′.*

Before we prove Proposition 1.2.1, we will establish a lemma that we will need for its proof, and which will also be useful for us later on. For this we need the *principle of uniform boundedness*, which states that if $(B_i, i \in \mathcal{I})$ is a family of operators in $\mathcal{L}(E)$ such that the sets $\{||B_i \psi||, i \in \mathcal{I}\}$ are bounded for each $\psi \in E$, then the set $\{||B_i||, i \in \mathcal{I}\}$ is also bounded.[5]

Lemma 1.2.2

1. *If $(T_t, t \geq 0)$ is a family of bounded linear operators on E so that $t \to T_t \psi$ is continuous for all $\psi \in E$, then for all $h > 0$,*

$$c_h = \sup\{||T_t||, 0 \leq t \leq h\} < \infty.$$

2. *If $(T_t, t \geq 0)$ is a family of bounded linear operators on E so that $t \to T_t \psi$ is continuous at zero for all $\psi \in E$, then there exists $h > 0$ so that $c_h < \infty$.*

---

[3] The reason for the nomenclature "C_0" is historical. The founders of the subject introduced a hierarchy of semigroups of type "C_j" (see Hille [45]). Only C_0 remains in general use today.
[4] A bounded linear operator X in E is a contraction if $||X|| \leq 1$, or equivalently $||X\psi|| \leq ||\psi||$ for all $\psi \in E$.
[5] This is proved in elementary texts on functional analysis, e.g., Simon [90], pp. 398–9.

Proof. 1. From the given continuity assumption, it follows that the set $\{||T_t\psi||, 0 \leq t \leq h\}$ is bounded, for all $\psi \in E$, and then the result follows from the principle of uniform boundedness.

2. Assume that the desired conclusion is false. Then $c_h = \infty$ for all $h > 0$. So taking $h = 1, 1/2, \ldots, 1/n, \ldots$, we can always find $0 < t_n < 1/n$ so that $||T_{t_n}|| > n$. But then by the uniform boundedness theorem, there exists $\psi \in E$ so that $\{||T_{t_n}\psi||, n \in \mathbb{N}\}$ is unbounded. But $\lim_{n\to\infty} T_{t_n}\psi = T_0\psi$, and this yields the required contradiction. $\square$

Note that in Lemma 1.2.2 (2), $c(h') \leq c(h) < \infty$ for all $0 \leq h' \leq h$, and if we assume that $T_0 = I$ therein, then $c(h') \geq 1$.

Proof of Proposition 1.2.1. Sufficiency is obvious. For necessity, let $t > 0$, $\psi \in E$ be arbitrary. Then for all $h > 0$, by (S1) and (S2)

$$||T_{t+h}\psi - T_t\psi|| \leq ||T_t||.||T_h\psi - \psi|| \to 0 \text{ as } h \to 0.$$

Then $t \to T_t\psi$ is right continuous from $[0, \infty)$ to E. To show left continuity, let h be as in Lemma 1.2.2 (2) so that $c_h < \infty$, and in view of the discussion after the proof of the last lemma, we take $h < t$. Since $t \to T_t\psi$ is continuous at zero, given any $\epsilon > 0$, there exists $\delta > 0$ so that if $0 < s < \delta$, then $||T_s\psi - \psi|| < \epsilon/[c_h^2(||T_{t-h}|| + 1)]$. Now choose $\delta' = \min\{\delta, h\}$. Then for all $0 < s < \delta'$,

$$\begin{aligned}
||T_t\psi - T_{t-s}\psi|| &\leq ||T_{t-s}||.||T_s\psi - \psi|| \\
&\leq ||T_{t-\delta'}||.||T_{\delta'-s}||.||T_s\psi - \psi|| \\
&< ||T_{t-\delta'}||c_{\delta'}\frac{\epsilon}{(c_h^2(||T_{t-h}|| + 1)} \\
&\leq \frac{||T_{t-h}||}{||T_{t-h}|| + 1}\frac{||T_{h-\delta'}||}{c_h}\frac{c_{\delta'}}{c_h}\epsilon \\
&< \frac{c_{h-\delta'}}{c_h}\epsilon \leq \epsilon
\end{aligned}$$

(where we have repeatedly used (S1)), and the proof is complete. $\square$

Example 1.2.3 Take $E = \mathbb{C}$. Fix $a \in \mathbb{C}$ and define for all $t \geq 0$,

$$T_t z = e^{ta}z,$$

for each z in $\mathbb{C}$. Then $(T_t, t \geq 0)$ is a C_0-semigroup and you can check that

- if $\Re(a) < 0$, then $(T_t, t \geq 0)$ is a contraction semigroup,
- if $\Re(a) = 0$, then T_t is an isometry for all $t > 0$,
- if $\Re(a) > 0$, then $\lim_{t\to\infty} ||T_t|| = \infty$.

Example 1.2.4 If A is a bounded operator on E, define

$$T_t = e^{tA} = \sum_{n=0}^{\infty} \frac{t^n}{n!} A^n.$$

You can check in Problem 1.1 that the series is norm convergent (uniformly on finite intervals) and that $(T_t, t \geq 0)$ is a C_0-semigroup. Note that

$$||e^{tA}|| \leq \sum_{n=0}^{\infty} \frac{t^n}{n!} ||A||^n = e^{t||A||}.$$

Example 1.2.5 (The Translation Semigroup) Here we take $E = C_0(\mathbb{R})$ or $L^p(\mathbb{R})$ for $1 \leq p < \infty$ and define

$$(T_t f)(x) = f(x + t).$$

Verifying (S1) and (S2) is trivial. We will establish (S3) in Chapter 3, when this example will be embedded within a more general class, or you can prove it directly in Problem 1.4.

Example 1.2.6 (Probabilistic Representations of Semigroups) Let $(\Omega, \mathcal{F}, P)$ be a probability space and $(X(t), t \geq 0)$ be a stochastic process taking values in $\mathbb{R}^d$. We obtain linear operators on $B_b(\mathbb{R}^d)$ by averaging over those paths of the process that start at some fixed point. To be precise, we define for each $f \in B_b(\mathbb{R}^d), x \in \mathbb{R}^d, t \geq 0$:

$$(T_t f)(x) = \mathbb{E}(f(X(t)) | X(0) = x).$$

Then we may ask when does $(T_t, t \geq 0)$ become a semigroup, perhaps on a nice closed subspace of $B_b(\mathbb{R}^d)$, such as $C_0(\mathbb{R}^d)$? We will see in Chapter 7 that this is intimately related to the *Markov property*.

Example 1.2.7 (Semidynamical Systems) Let M be a locally compact space and $(\tau_t, t \geq 0)$ be a semigroup of transformations of M so that $\tau_{s+t} = \tau_s \tau_t$ for all $s, t \geq 0$, τ_0 is the identity mapping and $t \rightarrow \tau_t x$ is continuous for all $x \in M$. Then we get a semigroup on $C_0(M)$ by the prescription

$$T_t f = f \circ \tau_t.$$

If $(\tau_t, t \geq 0)$ extends to a group $(\tau_t, t \in \mathbb{R})$, then we have reversible dynamics, since the system can be run both backwards and forwards in time. If this is not the case, we have irreversible dynamics. In physics, the latter is associated with entropy production, which yields the "arrow of time". These themes will be developed in Chapter 8.

For later work, the following general norm-estimate on C_0-semigroups will be invaluable.

Theorem 1.2.8 *For any C_0-semigroup $(T_t, t \geq 0)$, there exists $M \geq 1$ and $a \in \mathbb{R}$ so that for all $t \geq 0$,*

$$||T_t|| \leq M e^{at}. \tag{1.2.4}$$

Proof. First fix $n \in \mathbb{N}$, and define $c_n := \sup\{||T_t||, 0 \leq t \leq 1/n\} < \infty$ by Lemma 1.2.2(1). Since $||T_0|| = 1$, we must have $c_n \geq 1$. By repeated use of (S1), we deduce that $||T_t|| \leq c$, where $c := (c_n)^n$, for all $0 \leq t \leq 1$. Now for arbitrary $t \geq 0$, let $[t]$ denote the integer part of t. Again by repeated use of (S1), we have

$$\begin{aligned}
||T_t|| &\leq c^{[t]} c \\
&\leq c^{t+1} \\
&\leq M e^{at},
\end{aligned}$$

where $M := c$ and $a := \log(c)$. $\qquad\qquad\qquad\qquad\qquad\qquad\square$

From the form of M and a obtained in the proof of Theorem 1.2.8, we see that if $(T_t, t \geq 0)$ is a contraction semigroup, then $M = 1$ and $a = 0$. But in some cases it is possible that we may also find that there exists $b > 0$ so that $||T_t|| \leq e^{-bt}$, for all $t \geq 0$, due to some additional information. Indeed (see, e.g., Engel and Nagel [32] p. 5), one may introduce the *growth type $a_0 \geq -\infty$* of the semigroup:

$$a_0 := \inf\{a \in \mathbb{R}; \text{there exists } M_a \geq 1 \text{ so that } ||T_t|| \leq M_a e^{at} \text{ for all } t \geq 0\},$$

but we will not pursue this direction further herein.

For readers who like abstract mathematics, an (algebraic) *semigroup* is a set S that is equipped with a binary operation $*$, which is associative, i.e., $(a*b)*c = a*(b*c)$ for all $a, b, c \in S$. The semigroup is said to be a *monoid* if there is a neutral element e, so that $a*e = e*a = a$ for all $a \in S$. We have a *topological semigroup* (or monoid) if the set S is equipped with a topology such that the mapping $(a, b) \to a*b$ from $S \times S$ to S is continuous. Finally a *representation* of a topological monoid S is a mapping $\pi: S \to \mathcal{L}(E)$, where E is a real or complex Banach space so that $\pi(a*b) = \pi(a)\pi(b)$ for all $a, b \in S$, $\pi(e) = I$ and for all $\psi \in E$, the mapping $a \to \pi(a)\psi$ is continuous from S to E. Now $[0, \infty)$ is easily seen to be a topological monoid, under addition, when equipped with the usual topology inherited from $\mathbb{R}$, and what we have called a C_0-semigroup is simply a representation of $[0, \infty)$.

1.3 Unbounded Operators and Generators

1.3.1 Unbounded Operators and Density of Generators

Our next goal is to seek insight into the infinitesimal behaviour of semigroups. We will seek to answer the following question, motivated by our work in section 1.1: Is there always a linear operator A in E such that $A\psi = \frac{dT_t\psi}{dt}\big|_{t=0}$, and if so, what properties does it have? We will see that the answer to the first part of the problem is affirmative. But in general, we expect A to be an unbounded operator,[6] i.e., an operator that is only defined on a linear manifold D that is a proper subset of E; indeed this was exactly the case with the second-order differential operator that we considered in section 1.1. We use the term "unbounded" for such operators, since if $||A\psi|| \leq K||\psi||$ for all $\psi \in D$, and if D is dense in E, then it is easy to see that A may be extended to a bounded operator on the whole of E (see Problem 1.8(a)). The operators that we deal with will usually have the property that D is dense, but no such bounded extension as just discussed, can exist. Let us make a formal definition of linear operator that includes both the bounded and unbounded cases.

Let D be a linear manifold in a complex Banach space E. We say that $X : D \to E$ is a linear operator with domain D if

$$X(cf + g) = cXf + Xg$$

for all $f, g \in D$ and all $c \in \mathbb{C}$. We sometimes use the notation $\mathrm{Dom}(X)$ or D_X for the space D. The operator X is said to be *densely defined* if D is dense in E. We say that a linear operator X_1 having domain D_1 is an *extension* of X if $D \subseteq D_1$ and $X_1 f = Xf$ for all $f \in D$. In this case, we also say that X is a *restriction* of X_1 to D, and we write $X \subseteq X_1$.

Example 1.3.9 Let $E = C([0, 1])$ and consider the linear operator $(Xf)(x) = f'(x)$ for $0 \leq x \leq 1$, with domain $D = C^\infty([0, 1])$. Then X is densely defined. It has an extension to the space $D_1 = C^1([0, 1])$. Consider the sequence (g_n) in D where $g_n(x) = e^{-nx}$ for all $x \in [0, 1]$. Then $||Xg_n|| = n$, and so X is clearly unbounded.

Now let $(T_t, t \geq 0)$ be a C_0-semigroup acting in E. We define[7]

$$D_A = \left\{ \psi \in E; \exists \phi_\psi \in E \text{ such that } \lim_{t \to 0} \left\| \frac{T_t\psi - \psi}{t} - \phi_\psi \right\| = 0 \right\}.$$

---

[6] A good reference for unbounded operators, at least in Hilbert spaces, is Chapter VIII of Reed and Simon [76]. For Banach spaces, there is a great deal of useful information to be found in Yosida [101].

[7] The limit being taken here and in (1.3.5) below is one–sided.

It is easy to verify that D_A is a linear space and thus we may define a linear operator A in E, with domain D_A, by the prescription

$$A\psi = \phi_\psi,$$

so that, for each $\psi \in D_A$,

$$A\psi = \lim_{t \to 0} \frac{T_t\psi - \psi}{t}. \tag{1.3.5}$$

We call A the *infinitesimal generator*, or for simplicity, just the *generator*, of the semigroup $(T_t, t \geq 0)$. We also use the notation $T_t = e^{tA}$, to indicate that A is the infinitesimal generator of a C_0-semigroup $(T_t, t \geq 0)$. We have already seen examples, such as Example 1.2.4 above, where this notation agrees with familiar use of the exponential mapping. We will show below that D_A is dense in E. For now, in the general case, we may only assert that $0 \in D_A$.

Example 1.3.10 It is easy to see that the generators in Examples 1.1 and 1.2 are z with domain $\mathbb{C}$, and A with domain E (respectively). For Example 1.2.5 the generator is the differentiation operator $Xf = f'$, with domain

$$D_X := \{f \in C_0^1(\mathbb{R}); f' \in C_0(\mathbb{R})\}.$$

To see this, observe that for $f \in D_X$, given any $\epsilon > 0$, there exists $\delta > 0$ so that for all $|h| < \delta$

$$\sup_{x \in \mathbb{R}} \left| \frac{f(x+h) - f(x)}{h} - f'(x) \right| < \epsilon,$$

and the result follows. (Hint: First take $f \in C_c^\infty(\mathbb{R})$).

In order to explore properties of A we will need to use Banach space integrals. Let $0 \leq a < b < \infty$. We wish to integrate *continuous* functions Φ from $[a, b]$ to E. As we are assuming continuity, we can define $\int_a^b \Phi(s)ds$ to be an E-valued Riemann integral, rather than using more sophisticated techniques. To be precise, $\int_a^b \Phi(s)ds$ is the unique vector in E so that for any $\epsilon > 0$ there exists a partition $a = t_0 < t_1 < \cdots < t_n < t_{n+1} = b$ so that

$$\left\| \int_a^b \Phi(s)ds - \sum_{j=1}^{n+1} \Phi(u_j)(t_j - t_{j-1}) \right\| < \epsilon,$$

where $t_{j-1} < u_j < t_j$, for all $j = 1, \ldots, n+1$. The following basic properties will be used extensively in the sequel. They are all established by straightforward manipulations using partitions, or by variations on known arguments for the case $E = \mathbb{R}$ (see Problem 1.2).

Proposition 1.3.11 (Properties of the Riemann Integral) *Let* $\Phi \colon [0, \infty] \to E$ *be continuous.*

(RI1) For all $c > 0$, $\int_a^b \Phi(s + c)ds = \int_{a+c}^{b+c} \Phi(s)ds$.

(RI2) For all $a < c < b$, $\int_a^b \Phi(s)ds = \int_a^c \Phi(s)ds + \int_c^b \Phi(s)ds$.

(RI3) $\left\| \int_a^b \Phi(s)ds \right\| \leq \int_a^b \|\Phi(s)\|ds$.

(RI4) For all $t \geq 0$, $\lim_{h \to 0} \frac{1}{h} \int_t^{t+h} \Phi(s)ds = \Phi(t)$.

In addition, it is a straightforward exercise to check that for any $X \in \mathcal{L}(E)$

$$X \int_a^b \Phi(s)ds = \int_a^b X\Phi(s)ds. \tag{1.3.6}$$

We will mostly want to consider the case where $[a, b] = [0, t]$ and $\Phi(s) = T_s \psi$ for some fixed vector $\psi \in E$ and C_0-semigroup $(T_t, t \geq 0)$, so the desired continuity follows from (S3). From now on we will use the notation

$$\psi(t) := \int_0^t T_s \psi \, ds, \tag{1.3.7}$$

and we will frequently take $X = T_s$ for some $s > 0$ in (1.3.6) to obtain

$$T_s \psi(t) = \int_0^t T_{s+u} \psi \, du. \tag{1.3.8}$$

The following technical lemma will be very useful for us. In particular, it tells us that D_A contains much more than just the zero vector.

Lemma 1.3.12 *For each* $t \geq 0$, $\psi \in E$, $\psi(t) \in D_A$ *and*

$$A\psi(t) = T_t \psi - \psi.$$

Proof. Using (1.3.8), (RI1), (RI2) and (RI4), we find that for each $t \geq 0$,

$$\lim_{h \to 0} \frac{1}{h} \left[T_h \psi(t) - \psi(t) \right] = \lim_{h \to 0} \left(\frac{1}{h} \int_0^t T_{h+u} \psi \, du - \frac{1}{h} \int_0^t T_u \psi \, du \right)$$

$$= \lim_{h \to 0} \left(\frac{1}{h} \int_h^{t+h} T_u \psi \, du - \frac{1}{h} \int_0^t T_u \psi \, du \right)$$

$$= \lim_{h \to 0} \left(\frac{1}{h} \int_t^{t+h} T_u \psi \, du - \frac{1}{h} \int_0^h T_u \psi \, du \right)$$

$$= T_t \psi - \psi,$$

and the result follows. $\qquad\qquad\square$

We now show that the generator of a C_0-semigroup is always densely defined.

Theorem 1.3.13

(1) D_A *is dense in* E.
(2) $T_t D_A \subseteq D_A$ *for each* $t \geq 0$.
(3) $T_t A\psi = AT_t\psi$ *for each* $t \geq 0$, $\psi \in D_A$.

Proof. (1) By Lemma 1.3.12, $\psi(t) \in D_A$ for each $t \geq 0$, $\psi \in E$, but by (RI4), $\lim_{t\to 0}(\psi(t)/t) = \psi$; hence D_A is dense in E as required.

For (2) and (3), suppose that $\psi \in D_A$ and $t \geq 0$; then, by the definition of A and the continuity of T_t, we have

$$\left[\lim_{h\to 0} \frac{1}{h}(T_h - I)\right] T_t\psi = \lim_{h\to 0} \frac{1}{h}(T_{t+h} - T_t)\psi$$

$$= T_t \left[\lim_{h\to 0} \frac{1}{h}(T_h - I)\right]\psi = T_t A\psi.$$

From this it is clear that $T_t\psi \in D_A$ whenever $\psi \in D_A$, and so (2) is satisfied. We then obtain (3) when we take the limit. $\qquad\square$

1.3.2 Differential Equations in Banach Space

Let D be a dense linear manifold in E, I be an interval in $\mathbb{R}$ and $t \to \psi(t)$ be a mapping from I to D. Let $t \in I$ be such that there exists $\delta > 0$ so that $(t - \delta, t + \delta) \subseteq I$. We say that the mapping ψ is (strongly) differentiable at t if there exists $\psi'(t) \in E$ so that

$$\lim_{h\to 0} \left\| \frac{\psi(t+h) - \psi(t)}{h} - \psi'(t) \right\| = 0.$$

We then call $\psi'(t)$ the (strong) derivative of ψ at t, and (with the usual abuse of notation) we write $\frac{d\psi}{dt} := \psi'(t)$. If $O \subseteq I$ is open we say that ψ is differentiable on O if it is differentiable (in the above sense) at every point in O. By a standard argument, we can see that if ψ is differentiable on O, then it is continuous there. In principle, we could try to solve differential equations in Banach space that take the general form

$$\frac{d\psi}{dt} = F(\psi, t),$$

where $F: D \times O \to E$ is a suitably regular mapping. We are only going to pursue this theme in the special case where $I = [0, \infty)$, $O = (0, \infty)$, $D = D_A$ and $F(\psi, t) = A\psi$, where A is the generator of a C_0-semigroup. For more general investigations in this area, see, e.g., Deimling [28].

Lemma 1.3.14

1. *If $f \in D_A$ then for all $t \geq 0$,*

$$T_t f - f = \int_0^t T_s A f \, ds. \tag{1.3.9}$$

2. *The mapping $\psi(t) = T_t f$ is a solution of the initial value problem (ivp)*

$$\frac{d\psi}{dt} = A\psi, \quad \psi(0) = f.$$

Proof. 1. Let $\phi \in E'$ and define $F : [0, \infty) \to \mathbb{C}$ by

$$F(t) = \left\langle T_t f - f - \int_0^t T_s A f, \phi \right\rangle,$$

where we recall that $\langle \cdot, \cdot \rangle$ denotes the dual pairing between E and E'. Then the right derivative $D_+ F(t) = \langle A T_t f - T_t A f, \phi \rangle = 0$ for all $t > 0$, by Theorem 1.3.13(3) and (RI4). Since $F(0) = 0$ and F is continuous, it follows by a variation on the mean value theorem (see Lemma 1.4.4 on p. 24 of Davies [27] for details) that $F(t) = 0$ for all $t > 0$. Since ϕ is arbitrary, the result follows.

2. From (1), (RI2) and Theorem 1.3.13 (3), we have

$$\frac{1}{h}(T_{t+h} f - T_t f) = \frac{1}{h} \int_t^{t+h} A T_s f \, ds,$$

and the result follows when we pass to the limit using (RI4). $\qquad\square$

In relation to Lemma 1.3.14, we would like to be able to show that $u(t, \cdot) := T_t f$ is the *unique solution* to our ivp. We will do this in the next theorem.

Theorem 1.3.15 *If A is the generator of $(T_t, t \geq 0)$ and $\psi : [0, \infty) \to D_A$ is such that $\psi_0 = f$ and $\frac{d\psi}{dt} = A\psi$, then $\psi(t) = T_t f$ for all $t \geq 0$.*

Proof. Let $\phi \in E'$ and fix $t > 0$. For $0 \leq s \leq t$, define

$$F(s) := \langle T_t \psi(t - s), \phi \rangle.$$

Then the right derivative

$$D_+ F(s) = \lim_{h \to 0} \left\langle \frac{1}{h}(T_{s+h} \psi(t - s - h) - T_s \psi(t - s)), \phi \right\rangle$$

$$= \lim_{h \to 0} \left\langle \frac{1}{h}(T_{s+h} - T_s)\psi(t - s - h), \phi \right\rangle$$

$$+ \lim_{h \to 0} \left\langle \frac{1}{h}(T_s(\psi(t - s - h) - \psi(t - s)), \phi \right\rangle$$

$$= \langle A T_s \psi(t - s), \phi \rangle - \langle T_s A \psi(t - s), \phi \rangle$$
$$= 0,$$

by Theorem 1.3.13 (3). F is continuous, so as in the proof of Lemma 1.3.14, we have $F(t) = F(0)$ for all $t > 0$, i.e., $\langle T_t \psi_0, \phi \rangle = \langle \psi(t), \phi \rangle$, and so $\psi(t) = T_t \psi_0$. This shows that any solution of the ivp is generated from the initial value by a C_0-semigroup with generator A.

To complete the proof, suppose that $(T_t, t \geq 0)$ and $(S_t, t \geq 0)$ are two distinct C_0-semigroups having the same generator A. Then $\psi(t) = T_t \psi_0$ and $\xi(t) = S_t \psi_0$ both satisfy the conditions of the theorem. Then we conclude from the above discussion that $S_t \psi_0 = T_t \psi_0$ for all $\psi_0 \in D_A$. But D_A is dense in E and hence $S_t = T_t$ for all $t \geq 0$, and the result follows. $\qquad \square$

1.3.3 Generators as Closed Operators

Unbounded operators cannot be continuous, as they would then be bounded. The closest we can get to continuity is the property of being closed, which we now describe. Let X be a linear operator in E with domain D_X. Its *graph* is the set $G_X \subseteq E \times E$ defined by

$$G_X = \{(\psi, X\psi); \psi \in D_X\}.$$

We say that the operator X is *closed* if G_X is closed in $E \times E$. You can check that this is equivalent to the requirement that, for every sequence $(\psi_n, n \in \mathbb{N})$ in D_X which converges to $\psi \in E$, and for which $(X\psi_n, n \in \mathbb{N})$ converges to $\phi \in E$, $\psi \in D_X$ and $\phi = X\psi$. If X is closed and $D_X = E$, then the *closed graph theorem* states that X is bounded.

If X is a closed linear operator, then it is easy to check that its domain D_X is itself a Banach space with respect to the *graph norm* $|||\cdot|||$ where

$$|||\psi||| = ||\psi|| + ||X\psi||$$

for each $\psi \in D_X$ (see Problem 1.9).

It is not difficult to construct examples of linear operators that are densely defined, but not closed, or closed but not densely defined. We are fortunate that generators of C_0-semigroups satisfy both of these properties.

Theorem 1.3.16 *If A is the generator of a C_0-semigroup, then A is closed.*

Proof. Let $(\psi_n, n \in \mathbb{N})$ be a sequence in E such that $\psi_n \in D_A$ for all $n \in \mathbb{N}$, $\lim_{n \to \infty} \psi_n = \psi \in E$ and $\lim_{n \to \infty} A\psi_n = \phi \in E$. We must prove that $\psi \in D_A$ and $\phi = A\psi$.

First observe that, for each $t \geq 0$, by continuity, equation (1.3.9) and Theorem 1.3.13 (3),

$$T_t \psi - \psi = \lim_{n \to \infty} (T_t \psi_n - \psi_n)$$

$$= \lim_{n \to \infty} \int_0^t T_s A \psi_n \, ds$$

$$= \int_0^t T_s \phi \, ds, \tag{1.3.10}$$

where the passage to the limit in the last line is justified by the fact that

$$\left\| \int_0^t T_s A \psi_n \, ds - \int_0^t T_s \phi \, ds \right\| \leq \int_0^t \| T_s (A \psi_n - \phi) \| ds$$

$$\leq t M \| (A \psi_n - \phi) \| \to 0, \text{ as } n \to \infty,$$

with $M := \sup\{ \|T_s\|, 0 \leq s \leq t \} < \infty$ by Lemma 1.2.2. Now, by (RI4) applied to (1.3.10), we have

$$\lim_{t \to 0} \frac{1}{t} (T_t \psi - \psi) = \phi,$$

from which the required result follows. $\qquad \square$

1.3.4 Closures and Cores

This section may be omitted at first reading.

In many situations, a linear operator only fails to be closed because its domain is too small. To accommodate this we say that a linear operator X in E is *closable* if it has a closed extension $\tilde{X}$. Hence X is closable if and only if there exists a closed operator $\tilde{X}$ for which $\overline{G_X} \subseteq G_{\tilde{X}}$. Note that there is no reason why $\tilde{X}$ should be unique, and we define the *closure* $\overline{X}$ of a closable operator X to be its smallest closed extension (i.e., its domain is the intersection of the domains of all of its closed extensions), so that $\overline{X}$ is the closure of X if and only if the following hold:

1. $\overline{X}$ is a closed extension of X;
2. if X_1 is any other closed extension of X, then $D_{\overline{X}} \subseteq D_{X_1}$.

The next theorem gives a useful practical criterion for establishing closability.

Theorem 1.3.17 *A linear operator X in E with domain D_X is closable if and only if for every sequence $(\psi_n, n \in \mathbb{N})$ in D_X which converges to 0 and for which $(X\psi_n, n \in \mathbb{N})$ converges to some $\phi \in E$, we always have $\phi = 0$.*

Proof. If X is closable, then the result is immediate from the definition. Conversely, let (x, y_1) and (x, y_2) be two points in $\overline{G_X}$. Our first task is to show that we always have $y_1 = y_2$. Let $(x_n^{(1)}, n \in \mathbb{N})$ and $(x_n^{(2)}, n \in \mathbb{N})$ be two sequences in D_X that converge to x; then $(x_n^{(1)} - x_n^{(2)}, n \in \mathbb{N})$ converges to 0 and $(X x_n^{(1)} - X x_n^{(2)}, n \in \mathbb{N})$ converges to $y_1 - y_2$. Hence $y_1 = y_2$ by the criterion.

From now on, we write $y = y_1 = y_2$ and define $X_1 x = y$. Then X_1 is a well-defined linear operator with

$$D_{X_1} = \{x \in E; \text{ there exists } y \in E \text{ such that } (x, y) \in \overline{G_X}\}.$$

Clearly X_1 extends X and by construction we have $G_{X_1} = \overline{G_X}$, so that X_1 is closed, as required. $\qquad\qquad\square$

It is clear that the operator X_1 constructed in the proof of Theorem 1.3.17 is the closure of X. Indeed, from the proof of Theorem 1.3.17, we see that a linear operator X is closable if and only if it has an extension X_1 for which

$$G_{X_1} = \overline{G_X}.$$

Having dealt with the case where the domain is too small, we should also consider the case where we know that an operator X is closed, but the domain is too large or complicated for us to work in it with ease. In that case it is very useful to have a core available.

Let X be a closed linear operator in E with domain D_X. A linear subspace C of D_X is said to be a *core* for X if

$$\overline{X|_C} = X,$$

i.e., given any $\psi \in D_X$, there exists a sequence $(\psi_n, n \in \mathbb{N})$ in C such that $\lim_{n \to \infty} \psi_n = \psi$ and $\lim_{n \to \infty} X \psi_n = X \psi$.

Now we return to the study of C_0-semigroups $(T_t, t \geq 0)$. The next result is extremely useful in applications.

Theorem 1.3.18 *If $D \subseteq D_A$ is such that*

1. *D is dense in E,*
2. *$T_t(D) \subseteq D$ for all $t \geq 0$,*

then D is a core for A.

Proof. Let $\overline{D}$ be the closure of D in D_A with respect to the graph norm $|||.|||$ (where we recall that $|||\psi||| = ||\psi|| + ||A\psi||$ for each $\psi \in D_A$).

Let $\psi \in D_A$; then by hypothesis (1), we know there exists $(\psi_n, n \in \mathbb{N})$ in D such that $\lim_{n \to \infty} \psi_n = \psi$. Define $\psi(t) = \int_0^t T_s \psi \, ds$ and $\psi_n(t) = \int_0^t T_s \psi_n \, ds$

for each $n \in \mathbb{N}$ and $t \geq 0$. Approximating $\psi_n(t)$ by Riemann sums, and using hypothesis (2), we deduce that $\psi_n(t) \in \overline{D}$. Now using (1.3.9), Lemma 1.3.12, Lemma 1.2.2 and (1.2.4) we find that there exists $C_t > 0$ so that

$$\lim_{n \to \infty} |||\psi(t) - \psi_n(t)|||$$

$$= \lim_{n \to \infty} ||\psi(t) - \psi_n(t)|| + \lim_{n \to \infty} ||A\psi(t) - A\psi_n(t)||$$

$$\leq \lim_{n \to \infty} \int_0^t ||T_s(\psi - \psi_n)||ds + \lim_{n \to \infty} ||(T_t\psi - T_t\psi_n|| + \lim_{n \to \infty} ||\psi - \psi_n)||$$

$$\leq (tC_t + Me^{at} + 1) \lim_{n \to \infty} ||\psi - \psi_n)|| = 0,$$

and so $\psi(t) \in \overline{D}$ for each $t \geq 0$.

Now using (1.3.9) again and also (RI4), we obtain

$$\lim_{t \to 0} \left|\left|\left|\frac{1}{t}\psi(t) - \psi\right|\right|\right|$$

$$= \lim_{t \to 0} \left|\left|\frac{1}{t}\int_0^t T_s\psi \, ds - \psi\right|\right| + \lim_{t \to 0} \left|\left|\frac{1}{t}A\psi(t) - A\psi\right|\right|$$

$$= \lim_{t \to 0} \left|\left|\frac{1}{t}\int_0^t T_s\psi \, ds - \psi\right|\right| + \lim_{t \to 0} \left|\left|\frac{1}{t}(T_t\psi - \psi) - A\psi\right|\right| = 0.$$

From this we can easily deduce that $D_A \subseteq \overline{D}$, from which it is clear that D is a core for A, as required. $\qquad\square$

Example 1.3.19 If we return to the translation semigroup discussed in Examples 1.3 and 1.7, then it is very easy to check the hypotheses of Theorem 1.3.18 and show that $C_c^\infty(\mathbb{R})$ is a core for the generator.

1.4 Norm-Continuous Semigroups

Let $(T_t, t \geq 0)$ be a family of linear operators in $\mathcal{L}(E)$ that satisfy (S1) and (S2), but instead of (S3), we have that for all $t \geq 0$, $T_s \to T_t$ in the norm topology in $\mathcal{L}(E)$ as $s \to t$, i.e.,

$$\lim_{s \to t} ||T_t - T_s|| = 0.$$

If we imitate the proof of Proposition 1.2.1, we find that the above convergence is equivalent to requiring $\lim_{s \to 0} ||T_s - I|| = 0$. Semigroups that satisfy this condition are said to be *norm continuous*. It is clear that every norm-continuous semigroup is a C_0-semigroup. Note that for a norm-continuous semigroup $(T_t, t \geq 0)$, the mapping $t \to T_t$ is continuous from $[0, \infty)$ to $\mathcal{L}(E)$, and so the (Riemann) integral $\int_0^t T_s ds$, is well defined in the sense discussed in section 1.1.

Example 1.4.20 If we return to Example 1.2.4, then using an $\epsilon/3$–argument, one can easily show that $e^{tA} := \sum_{n=0}^{\infty} \frac{t^n}{n!} A^n$ is norm continuous for $A \in \mathcal{L}(E)$ (see Problem 1.1(b)). The next result shows that this class of examples comprises the entirety of the norm-continuous semigroups.

Theorem 1.4.21 *A C_0-semigroup $(T_t, t \geq 0)$ is norm continuous if and only if its generator A is bounded.*

Proof. By the discussion in Example 1.4.20, we need only prove necessity. So let $(T_t, t \geq 0)$ be norm continuous. By (RI4) we can and will choose h sufficiently small so that

$$\left\| I - \frac{1}{h} \int_0^h T_t dt \right\| < 1.$$

Now for such a value of h, define $W = \int_0^h T_t dt$. Then W is bounded and invertible (and its bounded inverse is given in terms of a Neumann series[8]). Define $V \in \mathcal{L}(E)$ by $V := W^{-1}(T_h - I)$. We will show that V is the generator of the semigroup, and then we are done. First observe that for $t > 0$, by (1.3.8) and (RI2) we have

$$W(T_t - I) = \int_t^{t+h} T_s ds - \int_0^h T_s ds$$

$$= \int_h^{t+h} T_s ds - \int_0^t T_s ds$$

$$= (T_h - I) \int_0^t T_s ds.$$

Then by (1.3.6)

$$T_t - I = V \int_0^t T_s ds = \int_0^t V T_s ds,$$

and so

$$\frac{T_t - I}{t} = \frac{1}{t} \int_0^t V T_s ds.$$

---

[8] To be precise,

$$W^{-1} = \frac{1}{h} \sum_{n=0}^{\infty} \left(I - \frac{1}{h} W \right)^n.$$

From here, we can easily deduce that the mapping $t \rightarrow T_t$ is norm-differentiable and that

$$V = \lim_{t \to 0} \frac{T_t - I}{t}$$

is the generator. $\square$

1.5 The Resolvent of a Semigroup

1.5.1 The Resolvent of a Closed Operator

Let X be a linear operator in E with domain D_X. Its *resolvent set* is

$$\rho(X) := \{\lambda \in \mathbb{C}; \lambda I - X \text{ is a bijection from } D_X \text{ to } E\}.$$

The *spectrum* of X is the set $\sigma(X) = \rho(X)^c$. Note that every eigenvalue of X is an element of $\sigma(X)$. If $\lambda \in \rho(X)$, the linear operator $R_\lambda(X) = (\lambda I - X)^{-1}$ is called the *resolvent* of T. For simplicity, we will sometimes write $R_\lambda := R_\lambda(X)$, when there can be no doubt as to the identity of X.

We remark that $\lambda \in \rho(X)$ if and only if for all $g \in E$ there exists a unique $f \in D_X$ so that

$$(\lambda I - X)f = g.$$

If we take X to be a partial differential operator, as in section 1.1, we see that resolvents (when they exist) are the operators that generate unique solutions to *elliptic* equations. For the next result we recall from section 1.3.3 that the domain of a closed linear operator is a Banach space, when it is equipped with the graph norm.

Proposition 1.5.22 *If X is a closed linear operator in E with domain D_X and resolvent set $\rho(X)$, then, for all $\lambda \in \rho(X)$, $R_\lambda(X)$ is a bounded operator from E into D_X (where the latter space is equipped with the graph norm).*

Proof. We will need the inverse mapping theorem, which states that a continuous bijection between two Banach spaces always has a continuous inverse (see, e.g., Reed and Simon [76], p. 83). For each $\lambda \in \rho(X)$, $\psi \in D_X$,

$$||(\lambda I - X)\psi|| \leq |\lambda| \, ||\psi|| + ||X\psi|| \leq \max\{1, |\lambda|\} \, |||\psi|||.$$

So $\lambda I - X$ is bounded and hence continuous from D_X to E. The result then follows by the inverse mapping theorem. $\square$

It follows from Proposition 1.5.22 that if X is both densely defined and closed, then $R_\lambda(X)$ extends to an operator in $\mathcal{L}(E)$, which we continue to denote[9] as $R_\lambda(X)$. Indeed we have shown that for all $\psi \in D_X$ there exists $K > 0$ so that

$$\|R_\lambda(X)\psi\| \leq \||R_\lambda(X)\psi\|| \leq K\|\psi\|,$$

and the result follows by the density of D_X in E. It is a trivial, but useful, consequence of the definition of resolvent set that if $\lambda \in \rho(X)$, then for every $x \in D_X$ there exists $y \in E$ so that $x = R_\lambda(X)y$.

The next result summarises some key properties of resolvents:

Proposition 1.5.23 *Let X be a closed linear operator acting in E.*

1. *The resolvent set $\rho(X)$ is open.*
2. *For all $\lambda, \mu \in \rho(X)$,*

$$R_\lambda - R_\mu = (\mu - \lambda)R_\lambda R_\mu. \tag{1.5.11}$$

3. *For all $\lambda, \mu \in \rho(X)$,*

$$R_\lambda R_\mu = R_\mu R_\lambda.$$

4. *For all $\lambda \in \rho(X)$, $f \in D_X$,*

$$R_\lambda X f = X R_\lambda f.$$

Proof. 1. We will never use this directly, so we omit the proof and direct the reader to the literature (see, e.g., Lemma 8.1.3 in Davies [27] pp. 212–3).
2. For all $f \in E$, we have

$$(\lambda I - X)[R_\lambda - R_\mu - (\mu - \lambda)R_\lambda R_\mu]f$$
$$= f - (\lambda I - \lambda I + \mu I - X)R_\mu f$$
$$= 0,$$

but $\lambda I - X$ is invertible and so we must have $R_\lambda - R_\mu - (\mu - \lambda)R_\lambda R_\mu = 0$, as is required.
3. This follows immediately from (2).
4. First note that $X R_\lambda f$ is meaningful, as R_λ maps E to D_X. The result follows from writing

$$X R_\lambda f = -(\lambda I - X)(\lambda I - X)^{-1}f + \lambda R_\lambda f$$
$$= -(\lambda I - X)^{-1}(\lambda I - X)f + \lambda R_\lambda f$$
$$= R_\lambda X f. \qquad \square$$

---

[9] As is standard, we use the same notation, $R_\lambda(X)$, to denote the resolvent acting from E to E, and from D_X (equipped with the graph norm) to E; it should always be clear which of these we will mean from the context that we are in.

Although we will mostly be concerned with the case where X is an unbounded operator, the following result is instructive.

Proposition 1.5.24 *If X is a bounded linear operator in E and $\lambda \in \mathbb{C}$ with $|\lambda| > ||X||$, then $\lambda \in \rho_X$.*

Proof. It is well known from elementary Banach space theory that if $c \in \mathbb{C}$ such that $|c|.||X|| < 1$, then $I - cX$ is invertible. Hence, if $\lambda \in \mathbb{C}$ with $|\lambda| > ||X||$, then $\lambda^{-1}(\lambda I - X)$ is invertible, and the result follows. $\square$

An immediate consequence of the last result is that the spectrum $\sigma(X) \subseteq B_{||X||}(0) \subset \mathbb{C}$.

1.5.2 Properties of the Resolvent of a Semigroup

If $(T_t, t \geq 0)$ is a C_0-semigroup having generator A, then since A is closed and densely defined, $R_\lambda(A) = (\lambda I - A)^{-1}$ is a well-defined bounded linear operator in E for $\lambda \in \rho(A)$. We call it the *resolvent of the semigroup*. Of course, there is no a priori reason why $\rho(A)$ should be non-empty. The following key theorem will put that doubt to rest. Before we state it, we recall the key estimate (1.2.4) $||T_t|| \leq Me^{at}$ for some $M > 1, a \in \mathbb{R}$, for all $t \geq 0$. Using this estimate, it is not difficult to see that for any $h \in C((0, \infty))$ which satisfies $\int_0^\infty h(t)e^{at}dt < \infty$, we may define a bounded linear operator on E by the prescription

$$\left(\int_0^\infty h(t)T_t dt \right) \psi = \lim_{T \to \infty} \int_0^T h(t)T_t \psi dt, \qquad (1.5.12)$$

for all $\psi \in E$.

Theorem 1.5.25 *Let A be the generator of a C_0-semigroup $(T_t, t \geq 0)$ satisfying $||T_t|| \leq Me^{at}$ for all $t \geq 0$. The following hold:*

1. $\{\lambda \in \mathbb{C}; \Re(\lambda) > a\} \subseteq \rho(A)$,
2. for all $\Re(\lambda) > a$,

$$R_\lambda(A) = \int_0^\infty e^{-\lambda t} T_t dt, \qquad (1.5.13)$$

3. for all $\Re(\lambda) > a$,

$$||R_\lambda(A)|| \leq \frac{M}{\Re(\lambda) - a}. \qquad (1.5.14)$$

Proof. For each $\Re(\lambda) > a$, we define a linear operator $S_\lambda(A)$ on E by the Fourier–Laplace transform on the right-hand side of (1.5.13). Our goal is to prove that this really is the resolvent. Note first of all that $S_\lambda(A)$ is a bounded operator on E of the form (1.5.12); indeed, for each $\psi \in E, t \geq 0$, on using (RI3) and (1.2.4) we have,

$$||S_\lambda(A)\psi|| \leq \int_0^\infty e^{-\Re(\lambda)t}||T_t\psi||dt \leq ||\psi||M \int_0^\infty e^{(a-\Re(\lambda))t} dt$$

$$= \frac{M}{\Re(\lambda) - a}||\psi||.$$

Hence we have $||S_\lambda(A)|| \leq \dfrac{M}{\Re(\lambda) - a}$.

Now define $\psi_\lambda = S_\lambda(A)\psi$ for each $\psi \in E$. Then by (1.3.8), change of variable and (RI4), we have

$$\lim_{h\to0} \frac{1}{h}(T_h\psi_\lambda - \psi_\lambda)$$

$$= \lim_{h\to0} \left(\frac{1}{h} \int_0^\infty e^{-\lambda t}T_{t+h}\psi\, dt - \frac{1}{h} \int_0^\infty e^{-\lambda t}T_t\psi\, dt \right)$$

$$= \lim_{h\to0} \left(\frac{1}{h} \int_h^\infty e^{-\lambda(t-h)}T_t\psi\, dt - \frac{1}{h} \int_0^\infty e^{-\lambda t}T_t\psi\, dt \right)$$

$$= -\lim_{h\to0} e^{\lambda h}\frac{1}{h} \int_0^h e^{-\lambda t}T_t\psi\, dt + \lim_{h\to0} \frac{1}{h}(e^{\lambda h} - 1) \int_0^\infty e^{-\lambda t}T_t\psi\, dt$$

$$= -\psi + \lambda\psi_\lambda.$$

Hence $\psi_\lambda \in D_A$ and $A\psi_\lambda = -\psi + \lambda S_\lambda(A)\psi$, i.e., for all $\psi \in B$

$$(\lambda I - A)S_\lambda(A)\psi = \psi.$$

So $\lambda I - A$ is surjective for all $\lambda > 0$ and its right inverse is $S_\lambda(A)$.

Our proof is complete if we can show that $\lambda I - A$ is also injective. To establish this, assume that there exists $\psi \in D_A$ with $\psi \neq 0$ such that $(\lambda I - A)\psi = 0$, and define $\psi_t = e^{\lambda t}\psi$ for each $t \geq 0$. Then differentiation yields the initial-value problem

$$\psi_t' = \lambda e^{\lambda t}\psi = A\psi_t$$

with initial condition $\psi_0 = \psi$. But by Theorem 1.3.15, we have $\psi_t = T_t\psi$ for all $t \geq 0$. We then have

$$||T_t\psi|| = ||\psi_t|| = |e^{\lambda t}|\, ||\psi||,$$

and so $||T_t|| \geq ||T_t\psi||/||\psi|| = e^{\Re(\lambda)t}$. But this contradicts $||T_t|| \leq Me^{at}$, as is seen by taking $\Re(\lambda) > a + \frac{\log(M)}{t}$. Hence we must have $\psi = 0$, and the proof that $S_\lambda(A) = R_\lambda(A)$ is complete. This establishes (1) and (2), while (3) follows from the estimate for $S_\lambda(A)$, which was obtained at the beginning of this proof. □

Note that if $(T_t, t \geq 0)$ is a contraction semigroup, then it follows from Theorem 1.5.25 (1) that

$$\sigma(A) \subseteq (-\infty, 0] \times i\mathbb{R}.$$

If E is a Hilbert space and the semigroup is self-adjoint in that T_t is a self-adjoint operator for all $t \geq 0$, then $\sigma(A) \subseteq (-\infty, 0]$, and we may write the spectral decomposition as

$$T_t = \int_{-\sigma(A)} e^{-\lambda t} E(d\lambda),$$

for each $t \geq 0$, and the generator satisfies the "dissipativity condition"[10] $\langle Af, f \rangle \leq 0$, for all $f \in D_A$. This is why, in the literature, self-adjoint contraction semigroups are often written in the form $T_t = e^{-tB}$, where B is a positive, self-adjoint operator. We will study self-adjoint operators in greater depth in Chapter 4.

Since most semigroups encountered in applications are contraction semigroups, it is a natural question to ask why we bother with the more general C_0-class? One reason, is that the theory (as we are seeing) is very intellectually satisfying. Another reason is that norm-continuous semigroups, which are an important subclass, are not necessarily contractions. Finally if A is the generator of a contraction semigroup $(S_t, t \geq 0)$, then it is natural to want to consider the trivial perturbations $A_c = A + cI$ where $c \in \mathbb{R}$, having domain D_A. It is easy to see that A_c generates the semigroup $(T_t, t \geq 0)$ and that $T_t = e^{ct} S_t$ for all $t \geq 0$, but we can always find sufficiently large c so that T_t is not a contraction. In the next chapter, we will see that the solutions of PDEs may also give rise to C_0-semigroups that are not necessarily contractions.

1.6 Exercises for Chapter 1

1. (a) Let $A \in \mathcal{L}(E)$. Show that the series $\sum_{n=0}^{\infty} \frac{t^n}{n!} A^n$ is absolutely convergent (uniformly on finite intervals).
   (b) Deduce that $T_t = \sum_{n=0}^{\infty} \frac{t^n}{n!} A^n$ defines a norm-continuous semigroup.
   (c) If E is a Hilbert space and A is self–adjoint, show that the result of (b) agrees with that obtained by defining $T_t = e^{tA}$ using spectral theory.
2. Prove Proposition 1.3.11 and (1.3.6).

---

[10] We will say more about this, and generalise it to Banach spaces in the next chapter.

3. Let $(T_t, \geq 0)$ be a family of bounded linear operators in E for which there exists $M \geq 0$ such that $\sup_{0 \leq t \leq 1} ||T_t|| \leq M$. If (S3)$'$ is valid for these operators on a dense subset of E, show that it holds on the whole of E.

4. Prove directly that the translation semigroup of Example 1.2.5 is strongly continuous in both $C_0(\mathbb{R})$ and $L^p(\mathbb{R})$.
   [Hint: On L^p, find a suitable dense subspace and use the result of (3)].

5. Suppose that S is a bounded operator on E and that T is a closed operator having domain D_T. Show that $S + T$ and ST are both closed operators, having domain D_T. What can you say when T is only known to be closeable? What can you say about TS?

6. Suppose that $(T_t^{(1)}, t \geq 0)$ and $(T_t^{(2)}, t \geq 0)$ are C_0-semigroups in E for which

$$T_s^{(1)} T_t^{(2)} = T_t^{(2)} T_s^{(1)},$$

   for all $s, t \geq 0$.
   (a) Show that $(T_t^{(1)} T_t^{(2)}, t \geq 0)$ is a C_0-semigroup on E.
   (b) If for $i = 1, 2, (T_t^{(i)}, t \geq 0)$ has generator A_i, deduce that $(T_t^{(1)} T_t^{(2)}, t \geq 0)$ has generator $A_1 + A_2$.

7. If $c > 0$ and $(S_t, t \geq 0)$ is a C_0-semigroup with generator A, show that $T_t = e^{-ct} S_t$ also defines a C_0-semigroup. What is the generator of this semigroup? Can you express its domain in terms of that of A?

8. (a) If A is a densely defined linear operator in E such that there exists $K \geq 0$ so that $||A\psi|| \leq K ||\psi||$ for all $\psi \in D_A$, show that A has a unique bounded extension to a linear operator $\tilde{A}$ defined on the whole of E.
   (b) If A is densely defined and D_A is closed, show that $D_A = E$.
   (c) If A is the generator of a C_0-semigroup, show that D_A is closed if and only if the semigroup is norm-continuous.

9. Show that if X is a closed linear operator on a Banach space, then its domain is complete under the graph norm.

10. Let $(T_t, t \geq 0)$ be a C_0-semigroup for which $||T_t|| \leq M e^{at}$ for all $t \geq 0$, where $M \geq 1$ and $a \in \mathbb{R}$. Use induction to show that if $\Re(z) > a$ and $n \in \mathbb{N}$ then

$$R_z^n f = \int_0^\infty \frac{t^{n-1}}{(n-1)!} e^{-zt} T_t f \, dt.$$

2

The Generation of Semigroups

The purpose of the first part of this chapter is to establish necessary and sufficient conditions for a closed, densely defined linear operator to be the generator of a C_0-semigroup on a Banach space E. As was the case in Chapter 1, the presentation is strongly influenced by that of Davies [25, 27]. We begin by developing some useful tools.

2.1 Yosida Approximants

Throughout this section, A will be a closed densely defined linear operator on E, with domain D_A. We assume that there exists $a \in \mathbb{R}$ so that $(a, \infty) \times i\mathbb{R} \subseteq \rho(A)$. So by Theorem 1.5.25(1), the case where A is the generator of a C_0-semigroup satisfying $||T_t|| \leq Me^{at}$ for all $t \geq 0$ is included within our analysis. For each $\lambda \in \mathbb{R}$ with $\lambda > a$, define

$$A_\lambda = \lambda A R_\lambda(A). \tag{2.1.1}$$

We call A_λ a *Yosida approximant* to A. Note that since $R_\lambda(A)$ maps E to D_A, the operator is well-defined, linear and has domain E. We will soon see the sense in which it "approximates" A. Writing $R_\lambda := R_\lambda(A)$, as in Chapter 1, we have

Lemma 2.1.1 *For each $\lambda > a$,*

$$A R_\lambda = \lambda R_\lambda - I,$$

and so $A_\lambda \in \mathcal{L}(E)$.

Proof.

$$AR_\lambda = A(\lambda I - A)^{-1}$$
$$= (A - \lambda I)(\lambda I - A)^{-1} + \lambda R_\lambda$$
$$= \lambda R_\lambda - I,$$

and so A_λ, being the difference of two bounded operators, is itself bounded.
□

Lemma 2.1.2 $\lim_{\lambda \to \infty} ||AR_\lambda f|| = 0$ *for all* $f \in E$.

Proof. First let $f \in D_A$, then by Proposition 1.5.23 (4) and (1.5.14),

$$||AR_\lambda f|| = ||R_\lambda A f||$$
$$\leq \frac{M}{\lambda - a} ||Af||$$
$$\to 0 \text{ as } \lambda \to \infty.$$

The result then follows by a density argument, using the fact that, by Lemma 2.1.1 and (1.5.14) again,

$$||AR_\lambda|| = ||\lambda R_\lambda - I|| \leq 1 + \frac{\lambda M}{\lambda - a},$$

which is bounded as $\lambda \to \infty$.
□

Lemma 2.1.3 *For all* $f \in D_A$,

$$\lim_{\lambda \to \infty} A_\lambda f = Af.$$

Proof. By the definition of the resolvent, for all $b > a$, there exists $g \in E$ so that $f = R_b g$. Then using Proposition 1.5.23(2), we have for $\lambda > b$,

$$||A_\lambda f - Af||$$
$$= ||\lambda AR_\lambda R_b g - AR_b g||$$
$$= ||\lambda A(R_\lambda - R_b)(b - \lambda)^{-1} g - AR_b g||$$
$$= \left\|\left(\frac{\lambda}{\lambda - b} - 1\right) AR_b g - \frac{\lambda}{\lambda - b} AR_\lambda g\right\|$$
$$\leq \left(\frac{\lambda}{\lambda - b} - 1\right) ||AR_b g|| + \frac{\lambda}{\lambda - b} ||AR_\lambda g||$$
$$\to 0 \text{ as } \lambda \to \infty,$$

by Lemma 2.1.2.
□

Note that in the case where A is indeed the generator of a C_0-semigroup, Lemma 2.1.3 enables us to approximate it by generators of norm-continuous semigroups. It would be interesting to find conditions under which this limiting relation can be "integrated" and enable us to write "$T_t f = \lim_{\lambda \to \infty} e^{tA_\lambda} f$". These will be presented in the next theorem.

2.2 Classifying Generators

In this section, we will establish the key theorems that give necessary and sufficient conditions, firstly, for a closed densely defined linear operator A to generate a C_0-semigroup, and secondly a (slight generalisation of) a contraction semigroup. The Yosida approximant tools that we developed in the first part will play an important role in the proof of the first theorem.

Theorem 2.2.4 (Feller, Miyadera, Phillips) *A closed densely defined linear operator A is the generator of a C_0-semigroup $(T_t, t \geq 0)$ in E, satisfying the estimate $\|T_t\| \leq Me^{at}$ for all $t \geq 0$, if and only if*

(FMP1) $\{\lambda \in \mathbb{C}; \Re(\lambda) > a\} \subseteq \rho(A)$.
(FMP2)   *For all $\lambda > a$ and all $m \in \mathbb{N}$,*

$$\|R_\lambda(A)^m\| \leq \frac{M}{(\lambda - a)^m}.$$

Proof. First assume that A is indeed the generator of a semigroup as stated. Then (FMP1) holds, as was established in Theorem 1.5.25 (1). To establish (FMP2), let $f \in E$. Then by Theorem 1.5.25 (2) and (1.2.4) we have

$$\|R_\lambda(A)^m f\| = \left\| \int_0^\infty \cdots \int_0^\infty e^{-\lambda(t_1 + \cdots + t_m)} T_{t_1 + \cdots + t_m} f \, dt_1 \ldots dt_m \right\|$$

$$\leq M \int_0^\infty \cdots \int_0^\infty e^{-(\lambda - a)(t_1 + \cdots + t_m)} dt_1 \ldots dt_m \|f\|$$

$$\leq \frac{M}{(\lambda - a)^m} \|f\|,$$

and the result follows.

The hard part of the proof is to establish the converse. To that end, we introduce the Yosida approximants, $A_\lambda = \lambda A R_\lambda$, which make sense for all $\lambda > a$ since we are assuming (FMP1). By Lemma 2.1.1, each of these operators is bounded, and by Theorem 1.4.21, the prescription $T_t^\lambda = e^{tA_\lambda}$ for $t \geq 0$, gives rise to a norm-continuous semigroup. Now we estimate, using Lemma 2.1.1 and the assumption (FMP2),

$$\|T_t^\lambda\| = \|e^{t\lambda(-I + \lambda R_\lambda)}\|$$

$$\leq e^{-\lambda t} \sum_{n=0}^{\infty} t^n \frac{\lambda^{2n}}{n!} \|R_\lambda^n\|$$

$$\leq M e^{-\lambda t} \sum_{n=0}^{\infty} t^n \frac{\lambda^{2n}}{n!} (\lambda - a)^{-n}$$

$$= M e^{\frac{ta\lambda}{\lambda - a}}.$$

From here we can establish two useful facts:

(I) $\limsup_{\lambda \to \infty} \|T_t^\lambda\| \leq M e^{at}$.

(II) Writing $\frac{a\lambda}{\lambda - a} = a\left(1 + \frac{a}{\lambda - a}\right)$, we deduce that if $\lambda \geq 2a$, then

$$\|T_t^\lambda\| \leq M e^{2at}.$$

We next want to show that $T_t^\lambda f$ converges as $\lambda \to \infty$ (uniformly in bounded intervals of $[0, \infty)$), for all $f \in E$. First we show this for $f \in D_A$. Using (II) above, for all $0 \leq s \leq t$, we have for $\lambda, \mu \geq 2a$,

$$\left\| \frac{d}{ds}[T_{t-s}^\lambda T_s^\mu f] \right\| = \|T_{t-s}^\lambda(-A_\lambda + A_\mu)T_s^\mu f\|$$

$$= \|T_{t-s}^\lambda T_s^\mu(-A_\lambda + A_\mu)f\|$$

$$\leq M^2 e^{4at} \|(A_\mu - A_\lambda)f\|,$$

where the fact that $T_s^\mu A_\lambda = A_\lambda T_s^\mu$ is a consequence of Proposition 1.5.23 (3) and (4). Integrating both sides of the last inequality from 0 to t we find that

$$\|T_t^\lambda f - T_t^\mu f\| \leq t M^2 e^{4at} \|(A_\mu - A_\lambda)f\|.$$

The net $(A_\mu f, \mu \geq 2a)$ is Cauchy by Lemma 2.1.3, and this enables us to define linear operators on D_A by the prescription

$$T_t f = \lim_{\lambda \to \infty} T_t^\lambda f,$$

and the convergence is uniform on bounded intervals. To verify this last fact, observe that from the last inequality, if $0 \leq t \leq T$, and $a \geq 0$,

$$\|T_t^\lambda f - T_t^\mu f\| \leq T M^2 e^{4aT} \|(A_\mu - A_\lambda)f\|.$$

The case $a < 0$ is left to the reader.

Using (II) and a standard density argument, we see that T_t extends to a bounded operator on E for all $t \geq 0$. We will now show that $(T_t, t \geq 0)$ is a C_0-semigroup. It is clear that $T_0 = I$. Note that $||T_t|| \leq Me^{at}$ follows easily from (I). To see that $T_{s+t} = T_s T_t$ for all $s, t \geq 0$, first use the fact that for all $f \in D_A$,

$$||T_s T_t f - T_s^\lambda T_t^\lambda f|| \leq ||T_s||.||T_t f - T_t^\lambda f|| + ||T_t^\lambda||.||T_s f - T_s^\lambda f||,$$

and apply (II), before taking limits as $\lambda \to \infty$. Then an easy density argument extends this convergence to all $f \in E$.

To establish strong continuity, again begin by taking $f \in D_A$ and write

$$||T_t f - T_s f|| \leq ||T_t f - T_t^\lambda f|| + ||T_t^\lambda f - T_s^\lambda f|| + ||T_s^\lambda f - T_s f||;$$

then $\lim_{s \to t} ||T_t f - T_s f|| = 0$ follows by an $\epsilon/3$ argument, where the uniformity of convergence on finite intervals is needed to take care of the middle term. Finally we again extend this to all $f \in E$ using a density argument.

Let B be the generator of $(T_t, t \geq 0)$. The proof will be complete if we can show that $B = A$. We begin by using (1.3.9), so we have for all $f \in D_A, t \geq 0, \lambda > a$

$$T_t^\lambda f - f = \int_0^t T_s^\lambda A_\lambda f \, ds.$$

By (II) we have

$$||T_s A f - T_s^\lambda A_\lambda f|| \leq Me^{2as} ||A_\lambda f - A f|| + ||T_s^\lambda A f - T_s A f||,$$

and so, taking the limit as $\lambda \to \infty$, and using Lemma 2.1.3, we deduce that

$$T_t f - f = \int_0^t T_s A f \, ds.$$

Differentiating this last identity (in the strong sense), we find that $f \in D_B$ and $Bf = Af$, and so $A \subseteq B$. Now for all $\lambda > a$, the linear operators $\lambda I - A$ and $\lambda I - B$ are both one-to-one with range E, this follows for the first operator by (FMP1) and for the second because it generates the semigroup. Hence for all $f \in E$,

$$f = (\lambda I - A)R_\lambda(A)f = (\lambda I - B)R_\lambda(A)f,$$

so that $R_\lambda(A)f = R_\lambda(B)f$, from which it follows that $D_B = D_A$, and so $A = B$, as required. $\qquad \square$

Having done all the hard work in proving this important theorem, we can now reap the benefits, one of which is the corresponding result for contraction semigroups. In fact, it is convenient for applications to PDEs to work with a slightly larger class. We say that a C_0-semigroup $(T_t, t \geq 0)$ is *ω-contractive* if there exists $\omega \in \mathbb{R}$ so that for all $t \geq 0$,

$$\|T_t\| \leq e^{\omega t}.$$

So we have a contraction semigroup when $\omega \leq 0$.

Theorem 2.2.5 (Hille–Yosida) *Let A be a closed densely defined linear operator on E. The following are equivalent:*

(i) *A is the generator of an ω-contraction semigroup.*
(ii) *$(\omega, \infty) \subseteq \rho(A)$ and $\|R_\lambda(A)\| \leq \frac{1}{\lambda - \omega}$ for all $\lambda > \omega$.*
(iii) *The set $\{z \in \mathbb{C}; \Re(z) > \omega\} \subseteq \rho(A)$ and $\|R_z(A)\| \leq \frac{1}{\Re(z) - \omega}$ for all $\Re(z) > \omega$.*

Proof. (i) $\Rightarrow$ (iii) If $(T_t, \geq 0)$ is an ω-contraction semigroup, we have $M = 1$ and $a = \omega$ in (1.2.4). So (iii) is an immediate consequence of Theorem 2.2.4.

(iii) $\Rightarrow$ (ii) is obvious.

(ii) $\Rightarrow$ (i) This follows by employing Yosida approximants, exactly as in the second part of the proof of Theorem 2.2.4, and using the fact that for all $m \in \mathbb{N}, \lambda > \omega$

$$\|R_\lambda(A)^m\| \leq \|R_\lambda(A)\|^m \leq \frac{1}{(\lambda - \omega)^m}. \qquad \square$$

Note. In statements of the Hille–Yosida theorem for contraction semigroups, it is usual to take $\omega = 0$.

In the next section, we will look at applications of Theorem 2.2.5 to solving parabolic PDEs. Our final task in this section is to reformulate Theorem 2.2.5 so that we do not need the resolvent. To this end, let X be a linear operator acting in E with domain D_X and define

$$\mathcal{E}_X := \{(f, \phi) \in E \times E'; f \in D_X, \langle f, \phi \rangle = \|f\| = \|\phi\| = 1\}.$$

Then $\mathcal{E}_X \neq \emptyset$. To see this, let $f \in D_X$ with $\|f\| = 1$ be arbitrary, and V_f be the linear span of f. If $\psi \in V_f'$ is non-zero, then $\phi := \left(\frac{1}{\langle f, \psi \rangle}\right) \psi$ has the desired properties and extends to a bounded linear functional on E by the Hahn–Banach theorem.

We say that X is *dissipative* if $\Re\langle Xf, \phi \rangle \leq 0$ for all $(f, \phi) \in \mathcal{E}_X$. In a Hilbert space setting, this is equivalent to the requirement that $\Re\langle Xf, f \rangle \leq 0$, for all $f \in D_X$ (see Problem 2.2).

Theorem 2.2.6 (Lumer–Phillips 1) *Let A be a closed densely defined linear operator on E. The following are equivalent:*

(i) *A is the generator of a contraction semigroup.*
(ii) *A is dissipative and* $Ran(\lambda I - A) = E$ *for all* $\lambda > 0$.

Proof. (i) $\Rightarrow$ (ii). By the Hille–Yosida theorem, $\lambda \in \rho(A)$ for all $\lambda > 0$, and so $Ran(\lambda I - A) = E$. The operator A is dissipative since for $(f, \phi) \in \mathcal{E}_A$,

$$
\begin{aligned}
\Re\langle Af, \phi \rangle &= \Re \left(\lim_{h \to 0} \frac{1}{h} \langle T_h f - f, \phi \rangle \right) \\
&= \Re \left(\lim_{h \to 0} \frac{1}{h} \langle T_h f, \phi \rangle - 1 \right) \\
&\leq \lim_{h \to 0} \frac{1}{h} (\|T_h\| \cdot \|f\| \cdot \|\phi\| - 1) \\
&\leq 0,
\end{aligned}
$$

where we have used the fact that T_h is a contraction for all $h \geq 0$.

(ii) $\Rightarrow$ (i) If $(f, \phi) \in \mathcal{E}_A$, and $\lambda > 0$, then

$$
\begin{aligned}
\|(\lambda I - A)f\| &\geq |\langle (\lambda I - A)f, \phi \rangle| \\
&= |\lambda - \langle Af, \phi \rangle| \\
&\geq \lambda = \lambda \|f\|.
\end{aligned}
$$

This last inequality clearly extends (by scaling) to all $f \in D_A$, from which we deduce that the operator $\lambda I - A$ is injective. We are given, within assumption (ii), that it is surjective. Hence $\lambda \in \rho(A)$ and $R_\lambda(A)$ exists, and is bounded from E to D_A. So $R_\lambda(A)g \in D_A$ for each $g \in A$, and from the estimate which we have just derived, we obtain

$$
\|g\| \geq \lambda \|R_\lambda(A)g\|.
$$

Hence $\|R_\lambda(A)\| \leq 1/\lambda$, and the result follows from the Hille–Yosida theorem. $\square$

There is a modified version of the Lumer–Phillips theorem which is rather useful in applications. The conditions on A are weakened somewhat. We state the theorem below, but we do not give the proof. This can be found in Davies [27] p. 233.

Theorem 2.2.7 (Lumer–Phillips 2) *Let A be a closable densely defined linear operator on E, which satisfies the following two conditions:*

(a) There exists $\lambda > 0$ *so that* $Ran(\lambda I - A)$ *is dense in E.*
(b) For all $f \in D_A$, *there exists* $\phi \in E'$ *so that* $\|\phi\| = 1$, $\langle \phi, f \rangle = \|f\|$ *and*
$\Re\langle Af, \phi \rangle \leq 0$.

Then A is dissipative, and $\overline{A}$ *is the generator of a contraction semigroup.*

2.3 Applications to Parabolic PDEs

We return to the parabolic PDE we considered at the beginning of section 1.1. We begin with some formal considerations. The equation of interest is

$$\frac{\partial u}{\partial t} = Au, \tag{2.3.2}$$

where for $\psi \in C^\infty(\mathbb{R}^d)$,

$$A\psi(x) = \sum_{i,j=1}^d a_{ij}(x)\partial_{ij}^2 \psi(x) + \sum_{i=1}^d b_i(x)\partial_i \psi - c(x)\psi(x). \tag{2.3.3}$$

It is often more convenient to rewrite the operator A in "divergence form",

$$A\psi(x) = \sum_{i,j=1}^d \partial_i(a_{ij}(x)\partial_j)\psi(x) - \sum_{i=1}^d b_i'(x)\partial_i \psi - c(x)\psi(x), \tag{2.3.4}$$

where for each $1 \le i \le d$

$$b_i' = -b_i + \sum_{j=1}^d \partial_j a_{ji}.$$

We will consider the following boundary value problem (BVP) on a bounded open domain $U \subset \mathbb{R}^d$ with smooth boundary ∂U:

$$\frac{\partial u}{\partial t} = Au \text{ on } U \times [0, T]$$
$$u = 0 \text{ on } \partial U \times [0, T]$$
$$u(0, \cdot) = g \text{ on } U. \tag{2.3.5}$$

We will also find it useful to consider the associated *elliptic* boundary value problem:

$$Au = f \text{ on } U \times [0, T]$$
$$u = 0 \text{ on } \partial U \times [0, T], \tag{2.3.6}$$

for $f \in L^2(U)$.

We will assume that all coefficient functions c, b_i, a_{ij} and the initial condition g are smooth in $\overline{U}$. We also assume that the matrix-valued function $a = (a_{ij})$ is symmetric for each value of x and uniformly elliptic, in that there exists $K > 0$ so that

$$\inf_{x \in \mathbb{R}^d} a(x)\xi \cdot \xi \ge K|\xi|^2 \text{ for all } \xi \in \mathbb{R}^d.$$

We are going to review in as little detail as possible the theory of how to describe the solution to this BVP using semigroups. For those who want full details, we refer the reader to standard texts such as Evans [34], Pazy [74], Rennardy and Rogers [78], and Yosida [101]. We emphasise that our treatment will be somewhat sketchy. For this, you will need to know a little bit about Sobolev spaces. But in fact, just the definition of the norm and inner product that you can find in Appendix C, will be all that is needed here. For more details of the material that follows, we direct the reader to Evans [34], particularly sections 6.2 and 7.4.3.

2.3.1 Bilinear Forms, Weak Solutions and the Lax–Milgram Theorem

We return to the linear operator A of (2.3.4) and associate to it the bilinear form $B_0 : C_c^2(U) \times C_c^2(U) \to \mathbb{R}$ given by the prescription

$$B_0[u, v] = -\langle Au, v \rangle_{L^2(U)}. \tag{2.3.7}$$

You can easily check that $|B_0[u, v]| < \infty$ for each $u, v \in C_c^2(U)$, and a straightforward integration by parts yields $B_0[u, v] = B[u, v]$ where

$$B[u, v] = \int_U \left\{ \sum_{i,j=1}^d a_{ij}(x)\partial_i u(x)\partial_j v(x) \right.$$
$$\left. + \sum_{i=1}^d b_i'(x)\partial_i u(x)v(x) + c(x)u(x)v(x) \right\} dx. \tag{2.3.8}$$

The advantage of B over B_0 is a reduction in the order of differentiability required. In fact B is a bilinear form defined on $C_c^2(U)$ (technically speaking, it is an extension of the form B_0) and it is not difficult to check that the mapping defined by (2.3.8) extends by continuity to a bilinear form, which we continue to denote as B, from $H_0^1(U) \times H_0^1(U) \to \mathbb{R}$. In fact, this follows directly from (2.3.10) below.

We say that $u \in H_0^1(U)$ is a *weak solution* to the elliptic BVP (2.3.6) with $f \in L^2(U)$ if

$$B[u, v] = \langle f, v \rangle_{L^2(U)} \tag{2.3.9}$$

for all $v \in H_0^1(U)$. Our key tool for finding (unique) weak solutions is the following generalisation of the Riesz representation theorem:

Theorem 2.3.8 (The Lax–Milgram Theorem) *Let H be a real Hilbert space and $C : H \times H \to \mathbb{R}$ be a bilinear form that satisfies the following conditions*

(LM1) $|C(u, v)| \leq K_1 \|u\|.\|v\|$,
(LM2) $K_2 \|u\|^2 \leq C(u, u)$,

for all $u, v \in H$, where $K_1, K_2 > 0$. Then given any $f \in H^*$, there exists a unique $u \in H$ so that for all $v \in H$,

$$C(u, v) = \langle f, v \rangle.$$

H and H^* are often identified, since they are naturally isomorphic through the Riesz representation theorem (see Appendix E); but it is convenient for what follows to think of f as a bounded linear functional defined on H. We refer the reader to Evans [34], pp. 316–7 for the proof. You can also attempt this for yourself in Problem 2.8.

2.3.2 Energy Estimates and Weak Solutions to the Elliptic Problem

Taking $C = B$ and $H = H_0^1(U)$ in Theorem 2.3.8, we see that the key to finding a unique weak solution to (2.3.6) is to satisfy (LM1) and (LM2). The following *energy estimates* provide the key tools for achieving this. The first of these is obtained by repeatedly using the Cauchy–Schwarz inequality within (2.3.8) (see Problem 2.7). The second is a little trickier.

Theorem 2.3.9 *There exists $\alpha, \beta \geq 0$ and $\gamma > 0$ so that for all $u, v \in H_0^1(U)$,*

$$|B(u, v)| \leq \alpha \|u\|_{H_0^1(U)} \|v\|_{H_0^1(U)}, \tag{2.3.10}$$

$$\beta \|u\|^2_{H_0^1(U)} \leq B(u, u) + \gamma \|u\|^2_{L_0^2(U)}. \tag{2.3.11}$$

In fact (2.3.11) is a special case of a more general result known as *Gårding's inequality*. Notice that (2.3.10) immediately gives us (LM1), but using (2.3.11) to obtain (LM2) is a little tricky because of the term involving γ. To get around this problem, we consider the modified elliptic BVP in U given by

$$- Au + \mu u = f \text{ on } U \times [0, T]$$
$$u = 0 \text{ on } \partial U \times [0, T], \tag{2.3.12}$$

where $\mu \geq \gamma$. From the semigroup point of view, it is interesting to see the appearance of the operator $\mu I - A$ in (2.3.12). This will be exploited very soon. We will now obtain a unique weak solution to (2.3.12). First we need to introduce the modified bilinear form which belongs to it:

$$B_\mu[u, v] = B[u, v] + \mu \langle u, v \rangle_{L^2(U)},$$

for each $u, v \in H_0^1(U)$.

It is easy to see that B_μ satisfies (LM1). For (LM2), use (2.3.11) to obtain

$$B_\mu[u, u] = B[u, u] + \mu \|u\|_{L^2(U)}^2$$

$$\geq \beta \|u\|_{H_0^1(U)}^2 + (\mu - \gamma) \|u\|_{L^2(U)}^2$$

$$\geq \beta \|u\|_{H_0^1(U)}^2.$$

Now for any $f \in L^2(U)$, $\langle f, \cdot \rangle$ is a bounded linear functional on $L^2(U)$, and so it induces a bounded linear functional on $H_0^1(U)$; indeed we have for all $v \in H_0^1(U)$,

$$|\langle f, v \rangle| \leq \|f\| \cdot \|v\|_{L^2(U)} \leq \|f\| \cdot \|v\|_{H_0^1(U)},$$

and so by Theorem 2.3.8, the modified elliptic BVP has a unique weak solution for each choice of $\mu \geq \gamma$.

2.3.3 Semigroup Solution of the Parabolic Problem

We return to the parabolic BVP (2.3.5). It is a fact that we will not show here that there is a unique solution to the problem that lives in $H_0^1(U)$. We will show that this solution is described by the action of an ω-contractive semigroup. We will consider the linear operator A with domain $D_A = H^2(U) \cap H_0^1(U)$. The role of $H^2(U)$ is to allow all the differentiability we need, while $H_0^1(U)$ enables us to "bootstrap" from the weak solution. In fact, it is a consequence of a quite deep result – the *Friedrichs–Lax–Niremberg theorem* that any weak solution to (2.3.12) also lies in D_A.

We are going to apply Theorem 2.2.5. It is clear that A is densely defined. It is also closed, but we omit the proof as it depends on some lengthy computations. We will show that condition (ii) of Theorem 2.2.5 is satisfied with $\omega = \gamma$. First note that for each $\lambda > \gamma$, since u is the weak solution of the modified elliptic BVP (2.3.12), $u \in D_A$ and so for all $v \in H_0^1(U)$,

$$\langle (\lambda I - A)u, v \rangle = \langle f, v \rangle,$$

(where, for the rest of this section, all norms and inner products without a subscript are in $L^2(U)$). Then by density

$$(\lambda I - A)u = f,$$

and so, since u is unique, we deduce that $\lambda I - A$ is bijective, i.e., $(\gamma, \infty) \subseteq \rho(A)$ and $u = R_\lambda f$.

Furthermore for all $\lambda > \gamma$, again using the fact that u is the weak solution to (2.3.12), we have

$$B[u, u] = \langle f, u \rangle - \lambda ||u||^2$$
$$\leq ||f||.||u|| - \lambda ||u||^2.$$

But from the energy estimate (2.3.11), we have

$$0 \leq B[u, u] + \gamma ||u||^2$$
$$\leq ||f||.||u|| + (\gamma - \lambda)||u||^2.$$

Writing $u = R_\lambda f$, we then deduce that $||R_\lambda|| \leq \dfrac{1}{\lambda - \gamma}$, and we are done.

We have proved that A generates an ω-contractive semigroup in $L^2(U)$. We have not proved that the unique solution is obtained by applying the semigroup to the initial condition. Once we know that there is a unique solution, this follows from Theorem 1.3.15 and the fact that the semigroup preserves D_A.

We have outlined in this section how the solution of parabolic boundary value problems can be expressed in terms on C_0 semigroups. Remarkably, we can also do this for the hyperbolic problem:

$$\frac{\partial^2 u}{\partial t^2} = Au \text{ on } U \times [0, T]$$
$$u = 0 \text{ on } \partial U \times [0, T]$$
$$u(0, \cdot) = g_1, \frac{\partial u}{\partial t}(0, \cdot) = g_2 \text{ on } U, \qquad (2.3.13)$$

with the same conditions on the coefficients as considered for the parabolic problem. The trick is to rewrite the PDE as a parabolic vector–valued equation:

$$\frac{\partial \mathbf{v}}{\partial t} = \Gamma \mathbf{v}, \qquad (2.3.14)$$

where for all $t \geq 0, \mathbf{v}(t) = \begin{pmatrix} v_1(t) \\ v_2(t) \end{pmatrix}$, and we have introduced the functions $v_1 = u$ and $v_2 = \frac{\partial u}{\partial t}$. The operator-valued matrix $\Gamma := \begin{pmatrix} 0 & 1 \\ A & 0 \end{pmatrix}$ is the candidate to be the generator of our semigroup, and its domain needs careful consideration. We will just be content here by noting that when we formally compute the action of the matrix on the right-hand side of (2.3.14), we obtain

$$\frac{\partial v_1}{\partial t} = v_2, \quad \frac{\partial v_2}{\partial t} = A v_1,$$

and so $\frac{\partial^2 u}{\partial t^2} = Au$, as required. For a full exposition with detailed proofs, see e.g., Evans [34], pp. 443–5.

In the introduction to Chapter 1, we considered partial differential operators defined on the whole of $\mathbb{R}^d$, but the discussion of this chapter has for simplicity focussed on those defined on bounded regions in $\mathbb{R}^d$. For a treatment of the more general case, see, e.g., Engel and Nagel [31], pp. 411–9.

2.4 Exercises for Chapter 2

1. If $(T_t, t \geq 0)$ is a C_0-semigroup with generator A, and resolvent $R_\lambda(A)$ for $\lambda \in \rho(A)$, show that
   (a) for all $x \in E$,

   $$\lim_{\lambda \to \infty} \lambda R_\lambda(A)x = x,$$

   (b) for all $n \in \mathbb{N}$, D_{A^n} is dense in E. (Hint: use induction and the result of (a).)
2. Show that X is a dissipative operator in a Hilbert space H if and only if $\Re\langle Xf, f\rangle \leq 0$, for all $f \in D_X$.
3. Let X be a densely defined closed linear operator in a Hilbert space H. Prove that the following are equivalent.
   (a) X is dissipative.
   (b) $||(\lambda I - X)f|| \geq \Re(\lambda)||f||$ for all $f \in D_X, \lambda \in \mathbb{C}, \Re(\lambda) > 0$.
   (c) $||(\lambda I - X)f|| \geq \lambda ||f||$ for all $f \in D_X, \lambda > 0$.

   (*Taken from Sinha and Srivasta [91], pp. 54–5.*)
   [Hint: To prove (a) $\Rightarrow$ (b), begin by looking at $\Re\langle(X - \lambda I)f, f\rangle$, and then use the Cauchy–Schwarz inequality.]
4. Deduce that X is dissipative in a Hilbert space H if and only if for all $f \in D_X$,

   $$||(X + I)f|| \leq ||(X - I)f||.$$

5. A C_0-group is a family $(V_t, t \in \mathbb{R})$ of bounded linear operators in a Banach space E such that (S2) holds, (S1) holds for all $s, t \in \mathbb{R}$ and (S3) holds in the sense of continuity from $\mathbb{R}$ to $\mathcal{L}(E)$. The generator B of the group, and

its domain D_B are defined exactly as in (1.3.5) but with the limit now being two-sided.

(a) Show that V_t is invertible with $V_t^{-1} = V_{-t}$ for all $t \in \mathbb{R}$.

(b) Deduce that $(T_t^+, t \geq 0)$ and $(T_t^-, t \geq 0)$ are C_0-semigroups having generators B and $-B$ (respectively), where $T_t^+ = V_t$ for all $t \geq 0$ and $T_t^- = V_{-t}$ for all $t \geq 0$. Show also that for all $s, t \geq 0$,

$$T_s^+ T_t^- = T_t^- T_s^+.$$

(c) If $\|V_t\| \leq M e^{a|t|}$ for all $t \in \mathbb{R}$, where $a \in \mathbb{R}$ and $M \geq 1$, show that $\{\lambda \in \mathbb{C}; |\Re(\lambda)| > a\} \subseteq \rho(B)$ and for all $\lambda \in \mathbb{R}$ with $|\lambda| > a$ and $n \in \mathbb{N}$

$$\|R_\lambda(B)\|^n \leq \frac{M}{(|\lambda| - a)^n}.$$

[Hint: Use the result of (b) and the fact (which you should check) that $R_{-\lambda}(B) = -R_\lambda(-B)$ for all $\lambda \in \rho(-B) = -\rho(B).$]

6. Show that if B is a closed densely defined operator in E such that for all $\{\lambda \in \mathbb{C}; |\Re(\lambda)| > a\} \subseteq \rho(B)$ and for all $\lambda \in \mathbb{R}$ with $|\lambda| > a$ and $n \in \mathbb{N}$

$$\|R_\lambda(B)\|^n \leq \frac{M}{(|\lambda| - a)^n},$$

then B is the generator of a C_0-group $(V_t, t \in \mathbb{R})$ satisfying $\|V_t\| \leq M e^{a|t|}$ for all $t \in \mathbb{R}$. Here is a suggested strategy.

(a) Construct the Yosida approximants B_λ and $(-B)_\lambda$ for all $\lambda > a$ and argue as in the proof of Theorem 2.2.4 to deduce that $(T_t^+, t \geq 0)$ and $(T_t^-, t \geq 0)$ are C_0-semigroups, where, for all $t \geq 0$, $f \in E$,

$$T_t^+ f = \lim_{\lambda \to \infty} e^{tB_\lambda}, \quad T_t^- f = \lim_{\lambda \to \infty} e^{t(-B)_\lambda}.$$

(b) Deduce that for all $s, t \geq 0$,

$$T_s^+ T_t^- = T_t^- T_s^+.$$

(c) Show that $(T_t^+ T_t^-, t \geq 0)$ defines a C_0-semigroup whose generator is the zero operator. Hence show that T_t^+ is invertible with $T_t^- = (T_t^+)^{-1}$ for all $t \geq 0$.

(d) Conclude that the required C_0-group is given by

$$V_t = \begin{cases} T_t^+ & \text{if } t \geq 0 \\ T_{-t}^- & \text{if } t < 0. \end{cases}$$

This and the previous question establishes a *Feller–Miyadera–Phillips theorem* for C_0-groups. The presentation here is based on the account in Engel and Nagel [32], pp. 72–3.

7. Derive the energy inequality (2.3.10) directly from (2.3.8).
8. Prove the Lax–Milgram theorem as follows:
   (a) Fix $u \in H$. Show that $v \rightarrow C(u, v)$ is a bounded linear functional on H.
   (b) By the Riesz representation theorem, it follows from (a) that there exists a unique $w(u) \in H$ so that

   $$C(u, v) = \langle w(u), v \rangle.$$

   Show that the prescription $Au = w(u)$ defines a bounded linear operator A on H.
   (c) Use (LM1) and (LM2) to show that A is injective and has closed range R_A.
   (d) Use a proof by contradiction to show that $R_A = H$.
   (e) Deduce that given any $f \in H'$, there exists $u \in H$ such that for all $v \in H$,

   $$C(u, v) = \langle f, v \rangle.$$

   (f) Prove that u is unique.

3

Convolution Semigroups of Measures

In this chapter, we will make our first attempt at exploring the territory of Example 1.2.6. This will lead us to the study of convolution semigroups of probability measures, the beautiful Lévy–Khintchine formula that characterises these, and the representation of the associated semigroups and their generators as pseudo-differential operators. We begin with two important motivating cases.

3.1 Heat Kernels, Poisson Kernels, Processes and Fourier Transforms

3.1.1 The Gauss–Weierstrass Function and the Heat Equation

The *Gauss–Weierstrass function* is the mapping $\gamma : (0, \infty) \times \mathbb{R}^d \to (0, \infty)$ defined by

$$\gamma(t, x) := \frac{1}{(4\pi t)^{\frac{d}{2}}} \exp\left(-\frac{|x|^2}{4t}\right), \tag{3.1.1}$$

for all $t > 0$, $x \in \mathbb{R}^d$. Note that $|x|$ here denotes the Euclidean norm, $|x| := (x_1^2 + x_2^2 + \cdots + x_d^2)^{\frac{1}{2}}$, where $x = (x_1, x_2, \ldots, x_d)$. From now on we will use the notation $\gamma_t(x) := \gamma(t, x)$. The mapping $\widetilde{\gamma}$ from $(0, \infty) \times \mathbb{R}^d \times \mathbb{R}^d \to (0, \infty)$ which sends (t, x, y) to $\gamma_t(x - y)$ is called the *heat kernel*, for reasons that should become clearer below. Readers will hopefully recognise (3.1.1) as the formula for the density of a normal distribution having mean zero and variance $2t$. As we will soon see, the Gauss–Weierstrass function is intimately related to the *Laplace operator*, or *Laplacian*

$$\Delta := \sum_{i=1}^{d} \partial_i^2,$$

which we consider as an (unbounded) linear operator on $C_0(\mathbb{R}^d)$ with domain $C_c^2(\mathbb{R}^d)$. Some basic properties of the Gauss–Weierstrass function are contained in the following:

Proposition 3.1.1 *1. The mapping γ is C^∞ and satisfies the heat equation, i.e.,*

$$\frac{\partial \gamma}{\partial t} = \Delta \gamma.$$

2. The heat kernel is symmetric, i.e., $\tilde{\gamma}(t, x, y) = \tilde{\gamma}(t, y, x)$, for all $t > 0, x, y \in \mathbb{R}^d$.

3. For each $t > 0$, γ_t is a probability density function (pdf), i.e.,

$$\int_{\mathbb{R}^d} \gamma_t(x)dx = 1.$$

4. For all $r > 0$, $\lim_{t \to 0} \int_{\{|x|>r\}} \gamma_t(x)dx = 0$.

Proof. (1) to (3) are standard, easily verified, and have hopefully been seen by readers before. For (4), substitute $y = t^{-1/2}x$ to find that

$$\int_{\{|x|>r\}} \gamma_t(x)dx = \frac{1}{(4\pi)^{\frac{d}{2}}} \int_{\mathbb{R}^d} 1_{\{|y|>t^{-1/2}r\}} e^{-|y|^2/4} dy$$

$$\to 0 \text{ as } t \to 0,$$

by use of dominated convergence. □

Recall that if f and g are sufficiently regular, then we can make sense of their convolution $f * g$ as a mapping defined on $\mathbb{R}^d$, where for all $x \in \mathbb{R}^d$,

$$(f * g)(x) = \int_{\mathbb{R}^d} f(x - y)g(y)dy. \tag{3.1.2}$$

For example, we might take $f, g \in L^1(\mathbb{R}^d)$, in which case $f * g \in L^1(\mathbb{R}^d)$. It is easy to see that $f * g = g * f$, and if $h \in L^1(\mathbb{R}^d)$, then $(f * g) * h = f * (g * h)$. With a little more work, you can show that $L^1(\mathbb{R}^d)$ is a (commutative) Banach algebra with $*$ as the product operation. We are interested in the heat equation, and for each $f \in C_0(\mathbb{R}^d), t > 0$, we define

$$u(t, \cdot) = f * \gamma_t.$$

It is easy to see that the mapping u is well-defined, so that for all $x \in \mathbb{R}^d$,

$$u(t, x) = \int_{\mathbb{R}^d} f(y)\gamma_t(x - y)dy = \int_{\mathbb{R}^d} f(y)\tilde{\gamma}_t(x, y)dy.$$

Theorem 3.1.2

1. *The mapping u is C^∞ on $(0, \infty) \times \mathbb{R}^d$.*
2. *The mapping u satisfies the heat equation $\frac{\partial u}{\partial t} = \Delta u$, with $\lim_{t \to 0} u(t, x) = f(x)$, and the convergence is uniform in $x \in \mathbb{R}^d$.*
3. *For all $t > 0$, $u(t, \cdot) \in C_0(\mathbb{R}^d)$ and for all $y \in \mathbb{R}^d$,*

$$\inf_{x \in \mathbb{R}^d} f(x) \leq u(t, y) \leq \sup_{x \in \mathbb{R}^d} f(x).$$

Proof. 1. is an easy exercise in the use of dominated convergence.

2. Again by dominated convergence, check that

$$\frac{\partial u}{\partial t} - \Delta u = \int_{\mathbb{R}^d} \left(\frac{\partial \gamma_t(x - y)}{\partial t} - \Delta \gamma_t(x - y) \right) f(y) dy = 0,$$

by Proposition 3.1.1 (1). For the convergence, first observe that by uniform continuity of f, given any $\epsilon > 0$, there exists $\delta > 0$ so that

$$\sup_{|x-y| < \delta} |f(x) - f(y)| < \epsilon/2.$$

Then

$$|u(t, x) - f(x)| \leq \int_{|x-y| < \delta} \gamma_t(x - y) |f(y) - f(x)| dy$$

$$+ \int_{|x-y| \geq \delta} \gamma_t(x - y) |f(y) - f(x)| dy.$$

The first integral is majorised by

$$\sup_{|x-y| < \delta} |f(x) - f(y)| \int_{\mathbb{R}^d} \gamma_t(x - y) dy < \epsilon/2,$$

where we have used Proposition 3.1.1 (3). Now by Proposition 3.1.1 (4), we can find t_0 so that for all $0 < t < t_0$, we have

$$\int_{\{|x| > \delta/2\}} \gamma_t(x) dx < \frac{\epsilon}{4\|f\|_\infty}.$$

We then find that

$$\int_{|x-y| \geq \delta} \gamma_t(x - y) |f(y) - f(x)| dy \leq 2\|f\|_\infty \int_{\{|x| > \delta/2\}} \gamma_t(x) dx < \epsilon/2,$$

and the required uniform convergence follows.

3. By dominated convergence,

$$\lim_{|x| \to \infty} u(t, x) = \int_{\mathbb{R}^d} \lim_{|x| \to \infty} f(x - y) \gamma_t(y) dy = 0.$$

Hence $u_t(\cdot) \in C_0(\mathbb{R}^d)$ by (1).

The left-hand estimate is obtained by observing that

$$u(t, x) \geq \inf_{x \in \mathbb{R}^d} f(x) \int_{\mathbb{R}^d} \gamma_t(y) dy = \inf_{x \in \mathbb{R}^d} f(x),$$

and the right-hand one is found similarly.

□

Note that Theorem 3.1.2 tells us that the Gauss–Weierstrass function is the *fundamental solution* to the heat equation.

An important property of the Gauss–Weierstrass function is the fact that for all $s, t > 0$,

$$\gamma_s * \gamma_t = \gamma_{s+t}. \tag{3.1.3}$$

This is not difficult to prove directly, but for later developments it will be useful to introduce some Fourier transform techniques. Let $f \in L^1(\mathbb{R}^d)$ be a probability density function, so that $f(x) \geq 0$ and $\int_{\mathbb{R}^d} f(x) dx = 1$. We define its *characteristic function* $\Phi_f : \mathbb{R}^d \to \mathbb{C}$, by the prescription

$$\Phi_f(y) = \int_{\mathbb{R}^d} e^{ix \cdot y} f(x) dx, \tag{3.1.4}$$

for all $y \in \mathbb{R}^d$.

Note that, in terms of the normalisation that we will use later on, the characteristic function is related to the Fourier transform $\widehat{f}$ of the function f via the formula:

$$\Phi_f(y) = (2\pi)^{d/2} \widehat{f}(-y).$$

Let us, for notational simplicity write $\phi_t := \Phi_{\gamma_t}$ for $t > 0$. Then it is well-known, and easy to verify[1] that

$$\phi_t(y) = e^{-t|y|^2},$$

for each $y \in \mathbb{R}^d$. We then obtain (3.1.3) immediately from this result and the following:

Proposition 3.1.3 *Let* $f, f_i \in L^1(\mathbb{R}^d)$ *for* $i = 1, 2$. *Then* $f = f_1 * f_2$ *if and only if* $\Phi_f = \Phi_{f_1} \Phi_{f_2}$.

Proof. Sufficiency is an easy calculation. For necessity, we then see that f and $f_1 * f_2$ both have the same characteristic function. But the mapping $f \to \Phi_f$ is injective (see, e.g., Theorem 9.5.7 in Dudley [30], p. 238), and the result follows. □

---

[1] We give a quick proof of this result at the end of Appendix F, using Itô's formula. For an alternative, more elementary approach, see Problem 3.1.

We define a family of linear operators $(T_t, t \geq 0)$ on $C_0(\mathbb{R}^d)$ by $T_0 = I$, and for $t > 0$, $f \in C_0(\mathbb{R}^d)$,

$$T_t f = f * \gamma_t. \tag{3.1.5}$$

Since $T_t f = u(t, \cdot)$, the fact that T_t does indeed preserve the space $C_0(\mathbb{R}^d)$ follows from Theorem 3.1.2 (3).

Theorem 3.1.4 *The operators* $(T_t, t \geq 0)$ *form a contraction semigroup on* $C_0(\mathbb{R}^d)$ *whose infinitesimal generator is an extension of* Δ.

Proof. The semigroup property essentially follows from (3.1.3), since for $f \in C_0(\mathbb{R}^d)$, $s, t > 0$, we have

$$\begin{aligned}
T_s T_t f &= (f * \gamma_t) * \gamma_s \\
&= f * (\gamma_t * \gamma_s) \\
&= f * \gamma_{s+t} = T_{s+t} f.
\end{aligned}$$

The axiom (S3) follows from the uniform convergence in Theorem 3.1.2 (2). The contraction property follows from an argument similar to that used to obtain the right-hand side of Theorem 3.1.2 (3). Finally the fact that Δ extends to the infinitesimal generator follows from the heat equation. $\qquad\square$

Note that a consequence of the last theorem is that Δ is a closed operator on its maximal domain

$$D_\Delta = \{f \in C_0(\mathbb{R}^d); \Delta f \in C_0(\mathbb{R}^d)\}.$$

The material in the subsection has been strongly influenced by that in section 2.7 of Grigor'yan [41], which is a wonderful reference for the heat kernel on both Euclidean spaces and more general Riemannian manifolds.

3.1.2 Brownian Motion and Itô's Formula

There is another way at looking at the ideas that we've just encountered. Instead of putting the emphasis on the heat kernel, we instead focus on the paths of a stochastic process. To be precise, let $(\Omega, \mathcal{F}, P)$ be a probability space and $B = (B(t), t \geq 0)$ be a Brownian motion defined on that space and taking values in $\mathbb{R}^d$. This means that

(B1) The process B has *independent increments,* i.e., the random variables $B(t_1) - B(t_0)$, $B(t_2) - B(t_1)$, ..., $B(t_n) - B(t_{n-1})$ are independent for all $0 = t_0 < t_1 < \cdots < t_n, n \in \mathbb{N}$.

(B2) $P(B(0) = 0) = 1$.

(B3) For all $0 \leq s < t < \infty$, $B(t) - B(s)$ is normally distributed with mean 0 and covariance matrix $2(t - s)I$.

(B4) There exists $\Omega' \in \mathcal{F}$ with $P(\Omega') = 1$ so that the mapping $t \to B(t)\omega$ is continuous from $[0, \infty)$ to $\mathbb{R}^d$ for all $\omega \in \Omega'$.

By (B3), for all $0 \leq s < t < \infty$, $A \in \mathcal{B}(\mathbb{R}^d)$, we have

$$P(B(t) - B(s) \in A) = \int_A \gamma_{t-s}(x)dx,$$

and this is the key link between Brownian motion and the Gauss–Weierstrass function. Furthermore, by (3.1.5), for all $t > 0$, $f \in C_0(\mathbb{R}^d)$, $x \in \mathbb{R}^d$,

$$
\begin{aligned}
T_t f(x) &= (f * \gamma_t)(x) \\
&= \int_{\mathbb{R}^d} f(x + y)\gamma_t(y)dy \\
&= \mathbb{E}(f(x + B(t))),
\end{aligned}
\tag{3.1.6}
$$

and we have obtained a probabilistic representation of our semigroup as an average over all of the possible paths of Brownian motion (cf. Example 1.2.5). Note also that because of (B2), the formula in fact holds for all $t \geq 0$.

If you have encountered *Itô's formula* [2] before, then you can get probabilistic insight into why the generator of our semigroup is an extension of Δ. Indeed for $f \in C_0^2(\mathbb{R}^d)$, that celebrated formula tells us that for all $t > 0$,

$$df(B(t)) = \sum_{i=1}^d (\partial_i f)(B(t))dB_i(t) + \sum_{i=1}^d (\partial_i^2 f)(B(t))dt,$$

and so, replacing f by its translate $f(\cdot + x)$ for $x \in \mathbb{R}^d$, we get

$$
\begin{aligned}
&= f(B(t) + x) - f(x) \\
&= \int_0^t \sum_{i=1}^d (\partial_i f)(B(s) + x)dB_i(s) + \int_0^t (\Delta f)(B(s) + x)ds,
\end{aligned}
$$

where $B(t) = (B_1(t), B_2(t), \ldots, B_d(t))$. Taking expectations of both sides, and using the fact that the stochastic integral has mean zero, yields

$$\mathbb{E}(f(B(t) + x)) - f(x) = \int_0^t \mathbb{E}(\Delta f)(B(s) + x))ds,$$

i.e.,

$$T_t f(x) - f(x) = \int_0^t T_s \Delta f(x)ds,$$

---

[2] For background and proofs, see, e.g., Chapter 4 of Oksendal [71] pp. 43–63, or Chapter 8 of Steele [92] pp. 111–37.

which should be compared with the general formula (1.3.9). For a crash course in stochastic integration and a proof of Itô's formula, at least in the one-dimensional case, see Appendix F. To learn more about the fascinating process of Brownian motion, see, e.g., Schilling and Partzsch [87] or Mörters and Peres [69].

There are many ways in which the ideas on heat kernels and Brownian motion can be generalised. We just mention one that will be useful to us later on. We introduce a *drift vector* $m \in \mathbb{R}^d$ and a *diffusion matrix* $a \in M_d(R)$ which is a positive definite symmetric matrix (so all of its eigenvalues are real and positive). We now define a variant on the Gauss–Weierstrass kernel,

$$\gamma_t^{a,m}(x) := \frac{1}{\sqrt{(4\pi t)^d \det(a)}} \exp\left\{-\frac{1}{4t}(x - mt) \cdot a^{-1}(x - mt)\right\},$$

for all $x \in \mathbb{R}^d, t > 0$. This is a probability density function, in fact, that of a normal distribution with mean mt and covariance matrix $2ta$. We again have the convolution property $\gamma_s^{a,m} * \gamma_t^{a,m} = \gamma_{s+t}^{a,m}$, for $s, t > 0$, and so we obtain a semigroup $(T_t, t \geq 0)$ on $C_0(\mathbb{R}^d)$ by the prescription $T_t f = f * \gamma_t^{a,m}$ for all $t > 0, f \in C_0(\mathbb{R}^d)$ (with $T_0 = I$). For later reference, we give the characteristic function:

$$\Phi_{\gamma_t^{a,m}}(y) = \exp\{t(im \cdot y - ay \cdot y)\}, \qquad (3.1.7)$$

for all $y \in \mathbb{R}^d$. We can obtain a process in $\mathbb{R}^d$ that has density $\gamma_t^{a,m}$ for $t > 0$, by taking σ to be a "square root of a", i.e., a $d \times p$ matrix so that $\sigma\sigma^T = a$. Then the required process is *Brownian motion with drift*, whose value at t is $mt + \sigma B(t)$, where $B = (B(t), t \geq 0)$ is a Brownian motion taking values in $\mathbb{R}^p$.

3.1.3 The Cauchy Distribution, the Poisson Kernel and Laplace's Equation

A natural question to ask at this stage is how special is the Gauss–Weierstrass function? Are there other families of functions $(f_t, t > 0)$ that have similar properties? The answer is yes, and to see that we introduce the family of *Cauchy distributions* $c_t : \mathbb{R}^d \to (0, \infty)$ for $t > 0$ by the prescription:

$$c_t(x) = \frac{\Gamma((d+1)/2)}{\pi^{(d+1)/2}} \frac{t}{(t^2 + |x|^2)^{(d+1)/2}}, \qquad (3.1.8)$$

for $x \in \mathbb{R}^d$, where Γ is the usual Euler gamma function: $\Gamma(y) = \int_0^\infty x^{y-1} e^{-x} dx$, for $\Re(y) > 0$.

When $d = 1$, it is straightforward to check that c_t is a probability density function. For $d > 1$, the simplest approach is to first calculate the characteristic function (see references below), and then check that $\Phi_{c_t}(0) = 1$. The formula $\tilde{c}_t(x, y) = c_t(x - y)$ for $x, y \in \mathbb{R}^d$ defines the *Poisson kernel*.[3] The Poisson kernel is rightly celebrated as it leads to the *Poisson integral formula* for solving the *Dirichlet problem* on the upper half-space $[0, \infty) \times \mathbb{R}^d$: i.e., to find a (harmonic) function $u : [0, \infty) \times \mathbb{R}^d \to \mathbb{R}$ so that

$$\partial_t^2 u(t, x) + \Delta u(t, x) = 0,$$

for all $(t, x) \in (0, \infty) \times \mathbb{R}^d$, and $u(0, \cdot) = f$, where $f \in C_0(\mathbb{R}^d)$. The solution is

$$u(t, \cdot) = f * c_t,$$

so that for all $x \in \mathbb{R}^d$, $u(t, x) = \int_{\mathbb{R}^d} f(y) \tilde{c}_t(x, y) dy$. Details can be found in, e.g., Stein and Weiss [95] pp. 47–9. We give a semigroup interpretation to this result below. Just as in the case of the Gauss–Weierstrass kernel, we have

$$c_{s+t} = c_s * c_t, \tag{3.1.9}$$

for all $s, t > 0$. This can be proved directly, or by applying Proposition 3.1.3, and the fact that $\Phi_{c_t}(y) = e^{-t|y|}$ for all $t > 0$, $y \in \mathbb{R}^d$. This latter fact is straightforward to prove in the case $d = 1$, by using the theory of residues. We now sketch this procedure.

We consider the function $g_{t,y}(z) = \frac{t}{\pi} \frac{e^{iyz}}{t^2 + z^2}$ for $z \in \mathbb{C}$, and observe that $\Phi_{c_t}(y) = \int_{\mathbb{R}} g_{t,y}(x) dx$. The function $g_{t,y}$ is clearly holomorphic, except for poles at $z = it$ and $z = -it$. If $y > 0$, we integrate over the contour $(-R, R) \cup \Gamma_R$ where Γ_R is a semicircle of radius R in the upper-half plane. By the residue theorem,

$$\int_{-R}^R g_{t,y}(x) dx + \int_{\Gamma_R} g_{t,y}(z) dz = 2\pi i (\text{Residue at } z = it).$$

Now the residue at $z = it$ is $\lim_{z \to it} (z - it) g_{t,y}(z) = \frac{1}{2\pi i} e^{-ty}$, and the result follows by taking limits as $R \to \infty$, using the easily established fact that $\lim_{R \to \infty} \int_{\Gamma_R} |g_{t,y}(z)| dz = 0$. For $y < 0$, we proceed in the same way, except that we integrate over a contour in the lower-half plane (see Problem 3.2). For the general case in higher dimensions, see, e.g., Theorem 1.14 in Stein and Weiss [95], pp. 6–7 or Example 2.11 in Sato [84], pp. 11–12.

---

[3] In the harmonic analysis literature, it is standard to use y instead of t in (3.1.8).

We can also show that for all $r > 0$, $\lim_{t \to 0} \int_{\{|x| > r\}} c_t(x) dx = 0$. This is proved in the same way as the corresponding result for the Gauss–Weierstrass kernel, except that the correct change of variable this time is $y = x/t$. Using (3.1.9) and this last result, we can imitate the proofs of Theorems 3.1.2 (2) and 3.1.4, to show that we obtain a contraction semigroup on $C_0(\mathbb{R}^d)$ by the prescription

$$T_0 = I \text{ and } T_t = f * c_t \text{ for } t > 0,$$

for each $f \in C_0(\mathbb{R}^d)$. We will see later in this chapter that (on a suitable domain) the infinitesimal generator of this semigroup is $-(-\Delta)^{1/2}$. In an L^2–context we could define this square root of a positive operator by using spectral theory, but we will make sense of it using pseudo-differential operators. The point of introducing the generator now is to relate the semigroup back to Laplace's equation. Define $u(t, \cdot) = T_t f$. Then we have

$$\frac{\partial u}{\partial t} = -(-\Delta)^{1/2} u,$$

and so formal differentiation[4] yields

$$\frac{\partial^2 u}{\partial t^2} + (-\Delta)^{1/2} \frac{\partial u}{\partial t} = 0.$$

Combining the last two identities, we get Laplace's equation on the upper half-space:

$$\frac{\partial^2 u}{\partial t^2} + \Delta u = 0.$$

Finally, there is a process corresponding to Brownian motion, called the *Cauchy process* $(Y(t), t \geq 0)$. It has independent increments and starts at 0 (almost surely), but it no longer has (almost surely) continuous sample paths. It is connected to the Cauchy distribution by the fact that

$$P(Y(t) - Y(s) \in A) = \int_A c_{t-s}(x) dx,$$

for all $0 \leq s < t < \infty$, $A \in \mathcal{B}(\mathbb{R}^d)$. Then we once again have a probabilistic representation for the semigroup (and hence for solutions of the Dirichlet problem) given by

$$T_t f(x) = \mathbb{E}(f(Y(t) + x)),$$

for all $f \in C_0(\mathbb{R}^d)$, $x \in \mathbb{R}^d$.

---

[4] We abandon rigour here, as the calculation provides so much insight.

We close this section with a remark. Both the Gauss–Weierstrass and Poisson kernels yield examples of the following phenomenon. We have a semigroup $(T_t, t \geq 0)$ on $C_0(\mathbb{R}^d)$ that is expressed in terms of a positive integrable kernel $k_t \colon \mathbb{R}^d \to (0, \infty)$ which is C^∞. Hence even though f is only assumed to be continuous, the mapping $u(t, \cdot) = T_t f$ defined for all $x \in \mathbb{R}^d$ by

$$u(t, x) = T_t f(x) = \int_{\mathbb{R}^d} f(y) k_t(x - y) dy,$$

is C^∞, as can be seen by using dominated convergence. This is often described as the *smoothing* property of the semigroup.

3.2 Convolution of Measures and Weak Convergence

This section is a diversion from the theme of the previous two, but it is essential to provide the tools to go further. Readers with knowledge of these two topics should feel free to skip ahead.

3.2.1 Convolution of Measures

Let $f \colon \mathbb{R}^d \to \mathbb{R}$ be a Borel measurable function and let $s \colon \mathbb{R}^{2d} \to \mathbb{R}^d$ be the mapping $s(x, y) = x + y$ for $x, y \in \mathbb{R}^d$. Then s is continuous and so the mapping $(x, y) \to f(x + y) = (f \circ s)(x, y)$ is Borel measurable from $\mathbb{R}^{2d}$ to $\mathbb{R}$. In particular, if $A \in \mathcal{B}(\mathbb{R}^d)$, then $(x, y) \to \mathbf{1}_A(x + y)$ is Borel measurable. Hence if μ_1, μ_2 are finite measures on $(\mathbb{R}^d, \mathcal{B}(\mathbb{R}^d))$ then we may define a set function m by the prescription

$$m(A) = \int_{\mathbb{R}^d} \int_{\mathbb{R}^d} \mathbf{1}_A(x + y) \mu_1(dx) \mu_2(dy). \tag{3.2.10}$$

Clearly we have $m(A) \leq \mu_1(\mathbb{R}^d) \mu_2(\mathbb{R}^d)$. Writing $A - x := \{a - x; a \in A\}$ for all $x \in \mathbb{R}^d$, we can easily verify that $\mathbf{1}_A(x + y) = \mathbf{1}_{A-x}(y)$ for each $y \in \mathbb{R}^d$. By Fubini's theorem, each of the mappings $x \to \mu_2(A - x)$ and $y \to \mu_1(A - y)$ from $\mathbb{R}^d$ to $\mathbb{R}$ are Borel measurable, and

$$m(A) = \int_{\mathbb{R}^d} \mu_2(A - x) \mu_1(dx) = \int_{\mathbb{R}^d} \mu_1(A - y) \mu_2(dy).$$

The set function m defines a measure on $(\mathbb{R}^d, \mathcal{B}(\mathbb{R}^d))$. To see this, observe that it is obvious that $m(\emptyset) = 0$, and if $(A_n, n \in \mathbb{N})$ is a sequence of mutually disjoint sets in $\mathcal{B}(\mathbb{R}^d)$ then since $\mathbf{1}_{\bigcup_{n \in \mathbb{N}} A_n} = \sum_{n \in \mathbb{N}} \mathbf{1}_{A_n}$, a straightforward application of Fubini's theorem yields

$$m\left(\bigcup_{n\in\mathbb{N}} A_n\right) = \sum_{n\in\mathbb{N}} \int_{\mathbb{R}^d} \int_{\mathbb{R}^d} \mathbf{1}_{A_n}(x+y)\mu_1(dx)\mu_2(dy)$$

$$= \sum_{n\in\mathbb{N}} m(A_n),$$

as required.

We call m the *convolution* of μ_1 and μ_2, and henceforth we write $m :=$ $\mu_1 * \mu_2$. It is easy to check that

- Convolution is associative, i.e., $\mu_1*(\mu_2*\mu_3) = (\mu_1*\mu_2)*\mu_3$ (see Problem 3.3).
- Convolution is commutative, i.e., $\mu_1 * \mu_2 = \mu_2 * \mu_1$ (see Problem 3.3).
- If μ_1 and μ_2 are probability measures, then so is $\mu_1 * \mu_2$.
- If each of μ_1 and μ_2 are absolutely continuous with respect to Lebesgue measure, with Radon–Nikodym derivatives[5] $g_i = d\mu_i/dx$ for $i = 1, 2$, then $\mu_1 * \mu_2$ is also absolutely continuous with respect to Lebesgue measure, and its Radon–Nikodym derivative g is given by $g = g_1 * g_2$ (in the sense of the usual convolution of functions (3.1.2)) (see Problem 3.4).

Theorem 3.2.5 *Let m, μ_1 and μ_2 be finite measures on $(\mathbb{R}^d, \mathbb{B}(\mathbb{R}^d))$. Then $m = \mu_1 * \mu_2$ if and only if*

$$\int_{\mathbb{R}^d} f(x)m(dx) = \int_{\mathbb{R}^d} \int_{\mathbb{R}^d} f(x+y)\mu_1(dx)\mu_2(dy), \qquad (3.2.11)$$

for all $f \in B_b(\mathbb{R}^d)$.

Proof. (Necessity) Suppose that $m := \mu_1 * \mu_2$. Then (3.2.11) holds in the case $f = \mathbf{1}_A$ by (3.2.10). Hence by linearity, it extends to the case where f is a non-negative simple function, and by monotone convergence to f being a non-negative measurable function. Finally if $f \in B_b(\mathbb{R}^d)$, we write $f = f_+ - f_-$ where $f_+ := \max\{f, 0\}$ and $f_- := \max\{-f, 0\}$, so that f is a difference of bounded non-negative measurable functions. Then

$$\int_{\mathbb{R}^d} f(x)m(dx) = \int_{\mathbb{R}^d} f_+(x)m(dx) - \int_{\mathbb{R}^d} f_-(x)m(dx)$$

$$= \int_{\mathbb{R}^d} \int_{\mathbb{R}^d} f_+(x+y)\mu_1(dx)\mu_2(dy)$$

$$- \int_{\mathbb{R}^d} \int_{\mathbb{R}^d} f_-(x+y)\mu_1(dx)\mu_2(dy)$$

$$= \int_{\mathbb{R}^d} \int_{\mathbb{R}^d} f(x+y)\mu_1(dx)\mu_2(dy),$$

---

[5] See Appendix E for background on this concept.

where we have used the (easily established) fact that $(\tau_y f)_\pm = \tau_y f_\pm$, for all $y \in \mathbb{R}^d$.

(Sufficiency) If (3.2.11) holds, then we may just take $f = 1_A$ and observe that this uniquely defines the measure m by (3.2.10). □

We close this subsection with some brief comments about the characteristic function Φ_μ of a probability measure μ on $\mathbb{R}^d$. This is defined for all $y \in \mathbb{R}^d$ by

$$\Phi_\mu(y) = \int_{\mathbb{R}^d} e^{ix\cdot y} \mu(dx). \qquad (3.2.12)$$

This generalises (3.1.4), where μ was assumed to have a density. The following properties are easily verified.

- $|\Phi_\mu(y)| \le \Phi_\mu(0) = 1$, for all $y \in \mathbb{R}^d$.
- $\Phi_\mu(-y) = \overline{\Phi_\mu(y)}$, for all $y \in \mathbb{R}^d$.
- The mapping $y \to \Phi_\mu(y)$ is continuous from $\mathbb{R}^d$ to $\mathbb{C}$ (this is an easy application of dominated convergence).
- If μ_1 and μ_2 are probability measures on $\mathbb{R}^d$, then for all $y \in \mathbb{R}^d$,

$$\Phi_{\mu_1 * \mu_2}(y) = \Phi_{\mu_1}(y)\Phi_{\mu_2}(y). \qquad (3.2.13)$$

3.2.2 Weak Convergence

Let S be a locally compact separable space[6] and $\mathcal{B}(S)$ be its Borel σ-algebra. All measures μ considered in this subsection are defined on $(S, \mathcal{B}(S))$. If $(\mu_n, n \in \mathbb{N})$ is a sequence of finite measures, we say that it *converges weakly* to the finite measure μ if

$$\lim_{n\to\infty} \int_S f(x)\mu_n(dx) = \int_S f(x)\mu(dx), \qquad (3.2.14)$$

for all $f \in C_b(S)$. A good reference for weak convergence is Chapter 1 of Billingsley [13]. There you can find a proof of the following useful theorem.

Theorem 3.2.6 (Portmanteau theorem) *The following are equivalent:*

1. *The sequence $(\mu_n, n \in \mathbb{N})$ converges weakly to μ.*
2. *The limit (3.2.14) is valid for all bounded uniformly continuous functions on S.*
3. $\limsup_{n\to\infty} \mu_n(A) \le \mu(A)$ *for all closed sets A in S.*

---

[6] Take $S \subseteq \mathbb{R}^d$ if you have not yet encountered this concept.

4. $\liminf_{n \to \infty} \mu_n(B) \geq \mu(B)$ *for all open sets A in S.*

5. $\lim_{n \to \infty} \mu_n(C) = \mu(C)$ *for all* $C \in \mathcal{B}(S)$ *such that* $\mu(\partial C) = 0$, *where* ∂C *denotes the boundary of the set C.*

To relate weak convergence to more familiar functional analytic notions, recall that if E is a Banach space and E' is its dual, then a sequence $(x_n, n \in \mathbb{N})$ in E converges to x as $n \to \infty$ in the *weak topology* on E if $l(x_n) \to l(x)$ as $n \to \infty$, for all $l \in E'$. A sequence $(l_n, n \in \mathbb{N})$ in E' converges to l as $n \to \infty$ in the *weak-$*$ topology* on E' if $l_n(x)$ converges to $l(x)$ as $n \to \infty$, for all $x \in E$. If for simplicity we take S to be compact, and let $E = C(S)$, then E' is the space of all regular (signed) finite Borel measures on S, equipped with the total variation norm, (see, e.g., Theorem 7.3.5 of Cohn [22], pp. 220–1) and we see that what we have called *weak convergence* of measures is in fact weak-$*$ convergence from the Banach space perspective.

From now on in this subsection we take $S = \mathbb{R}^d$. Weak convergence fits well with Fourier transform, as the following shows:

Lemma 3.2.7 *If* $(\mu_n, n \in \mathbb{N})$ *is a sequence of probability measures converging weakly to* μ, *then the corresponding sequence of characteristic functions* $(\Phi_{\mu_n}, n \in \mathbb{N})$ *converges pointwise to* Φ_μ.

Proof. For each $y \in \mathbb{R}^d$, as $n \to \infty$,

$$
\begin{aligned}
\Phi_{\mu_n}(y) &= \int_{\mathbb{R}^d} e^{ix \cdot y} \mu_n(dx) \\
&= \int_{\mathbb{R}^d} \cos(x \cdot y) \mu_n(dx) + i \int_{\mathbb{R}^d} \sin(x \cdot y) \mu_n(dx) \\
&\to \int_{\mathbb{R}^d} \cos(x \cdot y) \mu(dx) + i \int_{\mathbb{R}^d} \sin(x \cdot y) \mu(dx) \\
&= \int_{\mathbb{R}^d} e^{ix \cdot y} \mu(dx) = \Phi_\mu(y). \qquad \square
\end{aligned}
$$

The last easy lemma has a powerful converse, which we state without proof. That can be found in e.g. Itô [51], pp. 87–9.

Theorem 3.2.8 (Glivenko's Theorem) *If* $(\mu_n, n \in \mathbb{N})$ *is a sequence of probability measures and* μ *is another probability measure such that the corresponding sequence of characteristic functions* $(\Phi_{\mu_n}, n \in \mathbb{N})$ *converges pointwise to* Φ_μ, *then* $(\mu_n, n \in \mathbb{N})$ *converges weakly to* μ.

The notion of weak convergence of measures fits well with convolution as the next result demonstrates.

Theorem 3.2.9 *If $(\mu_n^{(i)}, n \in \mathbb{N})$ are sequences of probability measures converging weakly to $\mu^{(i)}$ for $i = 1, 2$, then the sequence $(\mu_n^{(1)} * \mu_n^{(2)}, n \in \mathbb{N})$ converges weakly to $\mu^{(1)} * \mu^{(2)}$.*

Proof. This is a straightforward application of Glivenko's theorem. Using (3.2.13), we have for all $y \in \mathbb{R}^d$,

$$|\Phi_{\mu_n^{(1)} * \mu_n^{(2)}}(y) - \Phi_{\mu^{(1)} * \mu^{(2)}}(y)|$$
$$= |\Phi_{\mu_n^{(1)}}(y)\Phi_{\mu_n^{(2)}}(y) - \Phi_{\mu^{(1)}}(y)\Phi_{\mu^{(2)}}(y)|$$
$$\leq |\Phi_{\mu_n^{(1)}}(y) - \Phi_{\mu^{(1)}}(y)||\Phi_{\mu_n^{(2)}}(y)| + |\Phi_{\mu^{(1)}}(y)||\Phi_{\mu_n^{(2)}}(y) - \Phi_{\mu^{(2)}}(y)|$$
$$\leq |\Phi_{\mu_n^{(1)}}(y) - \Phi_{\mu^{(1)}}(y)| + |\Phi_{\mu_n^{(2)}}(y) - \Phi_{\mu^{(2)}}(y)|$$
$$\to 0 \text{ as } n \to \infty. \qquad \square$$

Let $\mathcal{P}(\mathbb{R}^d)$ denote the space of all probability measures on $\mathbb{R}^d$, and τ be the topology of weak convergence. Then $(\mathcal{P}(\mathbb{R}^d), \tau)$ is a *Polish space* in that it is separable, and there is an equivalent topology that is induced by a metric, with respect to which $\mathcal{P}(\mathbb{R}^d)$ is complete. For details, and construction of this *Lévy metric*, see, e.g., Itô [51], pp. 100–2.

3.3 Convolution Semigroups of Probability Measures

In each of the examples that we have considered in subsections 3.1.1 and 3.1.3, we have a family of probability measures $(\mu_t, t > 0)$ on $(\mathbb{R}^d, \mathcal{B}(\mathbb{R}^d))$ given in the Gauss–Weierstrass case by $\mu_t(dx) = \gamma_t(x)dx$, and in the Cauchy case by $\mu_t(dx) = c_t(x)dx$, so that the prescription $T_0 = I$, and for $t > 0$,

$$T_t f(x) = \int_{\mathbb{R}^d} f(x + y)\mu_t(dy), \qquad (3.3.15)$$

for $f \in C_0(\mathbb{R}^d)$, $x \in \mathbb{R}^d$ gives rise to a contraction semigroup. More generally, suppose that we are given a family $(\mu_t, t \geq 0)$ of probability measures[7] on $(\mathbb{R}^d, \mathcal{B}(\mathbb{R}^d))$. When do we get a contraction semigroup by the prescription (3.3.15)? Up to now, we have not included the case $t = 0$ directly within our measures, but it is convenient to do this. In fact, if we require $T_0 f = f$ in (3.3.15), we find that $\mu_0 = \delta_0$ where for all $x \in \mathbb{R}^d$, δ_x is the *Dirac mass* at x, i.e., for all $A \in \mathcal{B}(\mathbb{R}^d)$,

$$\delta_x(A) = \begin{cases} 1 & \text{if } x \in A, \\ 0 & \text{if } x \notin A. \end{cases}$$

---

[7] More generally, we could take finite measures, but then we would lose the link with probability theory, and that would be a pity.

The prescription $T_{s+t} = T_s T_t$ for all $s, t \geq 0$ yields

$$\int_{\mathbb{R}^d} f(x+y)\mu_{s+t}(dy) = \int_{\mathbb{R}^d} \int_{\mathbb{R}^d} f(x+y_1+y_2)\mu_s(dy_1)\mu_t(dy_2). \quad (3.3.16)$$

Putting $x = 0$, we get

$$\int_{\mathbb{R}^d} f(y)\mu_{s+t}(dy) = \int_{\mathbb{R}^d} \int_{\mathbb{R}^d} f(y_1+y_2)\mu_s(dy_1)\mu_t(dy_2), \quad (3.3.17)$$

and this is precisely the defining property for convolution of measures as discussed in the previous section, i.e., for all $s, t \geq 0$ we have

$$\mu_{s+t} = \mu_s * \mu_t.$$

Furthermore we can always recover (3.3.16) from (3.3.17) by replacing f with its translate by x. For strong continuity at zero, we have

$$\lim_{t \to 0} \int_{\mathbb{R}^d} f(y)\mu_t(dy) = f(0) = \int_{\mathbb{R}^d} f(y)\delta_0(dy),$$

and this is (essentially)[8] weak convergence of probability measures.

These observations lead to the following definition:

A family $(\mu_t, t \geq 0)$ of probability measures on $(\mathbb{R}^d, \mathcal{B}(\mathbb{R}^d))$ is said to be a *convolution semigroup* of probability measures if

(C1) $\mu_{s+t} = \mu_s * \mu_t$, for all $s, t \geq 0$.
(C2) $\mu_0 = \delta_0$.
(C3) $\lim_{t \to 0} \mu_t = \delta_0$ (in the sense of weak convergence).

It can be shown that in the presence of (C1) and (C2), (C3) is equivalent to weak continuity of probability measures, i.e., $\lim_{s \to t} \int_{\mathbb{R}^d} f(x)\mu_s(dx) = \int_{\mathbb{R}^d} f(x)\mu_t(dx)$, for all $f \in C_b(\mathbb{R}^d)$. Next we will show how to obtain semigroups of operators from semigroups of measures.

Theorem 3.3.10 *If $(\mu_t, t \geq 0)$ is a convolution semigroup, then $(T_t, t \geq 0)$, as given by (3.3.15), is a contraction semigroup on $C_0(\mathbb{R}^d)$.*

Proof. We first prove that T_t preserves the space $C_0(\mathbb{R}^d)$ for each $t \geq 0$. To show that if f is continuous then $T_t f$ is also continuous is an easy application of dominated convergence. Indeed if (x_n) is a sequence in $\mathbb{R}^d$ converging to x, then the estimate $|T_t f(x_n)| \leq \|f\|_\infty$ for each $n \in \mathbb{N}$ enables us to conclude that $\lim_{n \to \infty} T_t f(x_n) = T_t f(x)$, as required. A similar argument shows that $\lim_{|x| \to \infty} T_t f(x) = 0$, and so $T_t(C_0(\mathbb{R}^d)) \subseteq C_0(\mathbb{R}^d)$. We have already seen how to establish the semigroup property $T_{s+t} = T_s T_t$ for all $s, t \geq 0$. The

---

[8] We should really work on $C_b(\mathbb{R}^d)$ to have this, but the apparent distinction really does not matter in this case – see, e.g., the discussion in Itô [51] on pages 98–9, or from a more general viewpoint, Theorem 1.1.9 on p. 25 of Heyer [44], to go into this in more detail.

contraction property is easily verified from (3.3.15). To prove strong convergence at 0, we first need to establish the following:

Claim. If $(p_t, t \geq 0)$ is a family of probability measures on $\mathbb{R}^d$, with $p_0 = \delta_0$ that is weakly continuous at $t = 0$, then given any $\epsilon > 0$ and $r > 0$, there exists $t_0 > 0$ so that for all $0 \leq t \leq t_0$, $p_t(B_r(0)^c) < \epsilon$.

To prove the claim we follow Malliavin et al. [65], pp. 98–9, and choose $f \in C_0(\mathbb{R}^d)$ to have support in $B_r(0)$ with $0 \leq f \leq 1$ and $f(0) = 1 - \epsilon/2$. Then by the given weak continuity, there exists $t_0 > 0$ so that for all $0 \leq t \leq t_0$, $\left| \int_{\mathbb{R}^d} (f(y) - f(0)) p_t(dy) \right| < \epsilon/2$. It follows that

$$p_t(B_r(0)^c) = 1 - p_t(B_r(0))$$
$$\leq 1 - \int_{B_r(0)} f(y) p_t(dy)$$
$$= 1 - \int_{\mathbb{R}^d} f(y) p_t(dy)$$
$$= 1 - f(0) + \int_{\mathbb{R}^d} (f(0) - f(y)) p_t(dy)$$
$$< \epsilon/2 + \epsilon/2 = \epsilon.$$

That completes the proof of the claim. To now prove the required strong convergence, observe that the result is trivial if $f = 0$, so assume that $f \neq 0$, and use the claim that we have just established to deduce that for any $\epsilon > 0$ and any $r > 0$, there exists $t_0 > 0$ such that $0 \leq t < t_0 \Rightarrow \mu_t(B_r(0)^c) < \epsilon/(4\|f\|_\infty)$.

Since every $f \in C_0(\mathbb{R}^d)$ is uniformly continuous, we can find $\delta > 0$ such that $\sup_{x \in \mathbb{R}^d} |f(x+y) - f(x)| < \epsilon/2$ for all $y \in B_\delta(0)$.

Choosing $r = \delta$, we then find that, for all $0 \leq t \leq t_0$,

$$\|T_t f - f\| = \sup_{x \in \mathbb{R}^d} |T_t f(x) - f(x)|$$
$$\leq \int_{B_\delta(0)} \sup_{x \in \mathbb{R}^d} |f(x+y) - f(x)| \mu_t(dy)$$
$$+ \int_{B_\delta(0)^c} \sup_{x \in \mathbb{R}^d} |f(x+y) - f(x)| \mu_t(dy)$$
$$< \frac{\epsilon}{2} \mu_t(B_\delta(0)) + 2\|f\|_\infty \mu_t(B_\delta(0)^c) < \epsilon,$$

and the required result follows. $\qquad\square$

Observe that for each $t \geq 0$, T_t extends to $B_b(\mathbb{R}^d)$ and the family $(T_t, t \geq 0)$ satisfies (S1) and (S2) on that space. Interestingly, we can then recapture both the convolution semigroup of measures, and their characteristic functions, from the semigroup of operators. To be precise, for all $t \geq 0$, $A \in \mathcal{B}(\mathbb{R}^d)$, $y \in \mathbb{R}^d$,

$$T_t 1_A(0) = \mu_t(A), \tag{3.3.18}$$

$$T_t e_y(0) = \Phi_{\mu_t}(y) \tag{3.3.19}$$

where $e_y(x) = e^{ix \cdot y}$, for all $x \in \mathbb{R}^d$.

We next present an important example of a convolution semigroup of measures $(\mu_t, t \geq 0)$ that is not of the form discussed up to now, i.e., there exists no $f_t \in L^1(\mathbb{R}^d)$ so that $\mu_t(dx) = f_t(x)dx$, for any $t \in \mathbb{R}$. This is the *compound Poisson semigroup* associated to a pair (ρ, c), where ρ is a probability measure on $\mathbb{R}^d$ and $c > 0$. It is defined as follows, for $t \geq 0$,

$$\mu_t = \sum_{n=0}^{\infty} e^{-ct} \frac{(ct)^n}{n!} \rho^{*n}. \tag{3.3.20}$$

Here, for all $n \in \mathbb{N}$, ρ^{*n} is the n-fold convolution of the measure ρ with itself, $\rho^{*0} := \delta_0$, and the infinite series is to be interpreted in the "weak sense", i.e., for all $f \in B_b(\mathbb{R}^d)$,

$$\int_{\mathbb{R}^d} f(x)\mu_t(dx) = \sum_{n=0}^{\infty} e^{-ct} \frac{(ct)^n}{n!} \int_{\mathbb{R}^d} f(x)\rho^{*n}(dx).$$

Observe that even if the measure ρ is assumed to be absolutely continuous with respect to Lebesgue measure, having density g (say), so that for all $n \in \mathbb{N}$, $\rho^{*n}(dx) = g^{*n}(x)dx$, then if $A \in \mathcal{B}(\mathbb{R}^d)$ is any set of Lebesgue measure zero with $0 \in A$, then $\mu_t(A) = 1$, and so μ_t cannot be absolutely continuous with respect to Lebesgue measure.

To see that $(\mu_t, t \geq 0)$ really is a convolution semigroup of probability measures, we find that for all $s, t \geq 0$,

$$\mu_s * \mu_t = \sum_{m=0}^{\infty} \sum_{n=0}^{\infty} e^{-c(s+t)} \frac{(cs)^n}{n!} \frac{(ct)^m}{m!} \rho^{*(n+m)}$$

$$= \sum_{m=0}^{\infty} \sum_{n=0}^{\infty} e^{-c(s+t)} \frac{c^{m+n}}{(m+n)!} \binom{m+n}{n} s^n t^m \rho^{*(n+m)}$$

$$= \sum_{p=0}^{\infty} \sum_{k=0}^{p} e^{-c(s+t)} \frac{c^p}{p!} \binom{p}{k} s^k t^{p-k} \rho^{*p}$$

$$= \sum_{p=0}^{\infty} e^{-c(s+t)} \frac{[c(s+t)]^p}{p!} \rho^{*p} = \mu_{s+t}$$

and for all $f \in C_b(\mathbb{R}^d)$,

$$\int_{\mathbb{R}^d} f(x)\mu_t(dx) = f(0) + \sum_{n=1}^{\infty} e^{-ct}\frac{(ct)^n}{n!}\int_{\mathbb{R}^d} f(x)\rho^{*n}(dx) \to f(0),$$

as $t \to 0$, by dominated convergence.

It is of interest to calculate the characteristic function of the measure μ_t for $t \geq 0$. Using Fubini's theorem we obtain the following,

$$
\begin{aligned}
\Phi_{\mu_t}(y) &= \int_{\mathbb{R}^d} e^{ix \cdot y}\mu_t(dx) \\
&= \sum_{n=0}^{\infty} e^{-ct}\frac{(ct)^n}{n!}\int_{\mathbb{R}^d} e^{ix \cdot y}\rho^{*n}(dx) \\
&= \sum_{n=0}^{\infty} e^{-ct}\frac{(ct)^n}{n!}\int_{\mathbb{R}^d}\cdots\int_{\mathbb{R}^d} e^{i(x_1+\cdots+x_n)\cdot y}\rho(dx_1)\ldots\rho(dx_n) \\
&= \sum_{n=0}^{\infty} e^{-ct}\frac{(ct)^n}{n!}\left(\int_{\mathbb{R}^d} e^{ix \cdot y}\rho(dx)\right)^n \\
&= \sum_{n=0}^{\infty} e^{-ct}\frac{(ct)^n}{n!}\Phi_\rho(y)^n \\
&= \exp\left(ct(\Phi_\rho(y) - 1)\right) \\
&= \exp\left\{t\int_{\mathbb{R}^d}(e^{i(x \cdot y)} - 1)c\rho(dx)\right\},
\end{aligned}
$$

for each $y \in \mathbb{R}^d$.

For another interesting example, return to the translation semigroup of Example 1.2.5. It is easy to check that this fits into the general scheme of (3.3.15), where the convolution semigroup of measures is $\{\delta_t, t \geq 0\}$.

Proposition 3.3.11 *If $(\mu_t^{(1)}, t \geq 0)$ and $(\mu_t^{(2)}, t \geq 0)$ are convolution semigroups, then so is $(\mu_t, t \geq 0)$ where for each $t \geq 0$, $\mu_t := \mu_t^{(1)} * \mu_t^{(2)}$.*

Proof. Exercise. $\qquad\square$

Note that if we calculate the characteristic function of the measure μ_t appearing in Proposition 3.3.11, we get for all $y \in \mathbb{R}^d$,

$$\Phi_{\mu_t}(y) = \Phi_{\mu_t^{(1)}}(y)\Phi_{\mu_t^{(2)}}(y).$$

We apply this to the case where $(\mu_t^{(1)}, t \geq 0)$ corresponds to Brownian motion with drift as in (3.1.7) and $(\mu_t^{(2)}, t \geq 0)$ is the compound Poisson semigroup, as just discussed. We then obtain

$$\Phi_{\mu_t}(y) = \exp\left\{ t \left(im \cdot y - ay \cdot y + \int_{\mathbb{R}^d} (e^{i(x \cdot y)} - 1)c\rho(dx) \right) \right\}. \quad (3.3.21)$$

Its worth pointing out that, by the discussion in section 3.3, for $t > 0$, μ_t is absolutely continuous with respect to Lebesgue measure.

The structure revealed in (3.3.21) leads us nicely into the next topic.

3.4 The Lévy–Khintchine Formula

The next result is the key theorem which characterises convolution semigroups of measures via their characteristic functions. From now on we will use the simplifying notation $\Phi_t := \Phi_{\mu_t}$ for the characteristic function of a measure μ_t. We will see that the theorem finds a function η, whose value at $y \in \mathbb{R}^d$ is the generator of the contraction semigroup of complex numbers $(\Phi_t(y), t \geq 0)$. As we will see later on, the inverse Fourier transform will then lead us to important expressions for the generator of $(T_t, t \geq 0)$, via the theory of pseudo-differential operators.

Theorem 3.4.12 (Lévy–Khintchine) *A family* $(\mu_t, t \geq 0)$ *of probability measures on* $\mathbb{R}^d$ *is a convolution semigroup if and only if its characteristic function has the form*

$$\Phi_t(y) = e^{t\eta(y)},$$

for all $t \geq 0$, $y \in \mathbb{R}^d$, *where*

$$\eta(y) = im \cdot y - ay \cdot y$$
$$+ \int_{\mathbb{R}^d} \left(e^{ix \cdot y} - 1 - i\frac{x \cdot y}{1 + |x|^2} \right) v(dx), \quad (3.4.22)$$

where $m \in \mathbb{R}^d$, a *is a* $d \times d$ *non-negative definite symmetric matrix, and* v *is a Borel measure on* $\mathbb{R}^d$ *for which*

$$v(\{0\}) = 0 \text{ and } \int_{\mathbb{R}^d} (1 \wedge |x|^2)v(dx) < \infty. \quad (3.4.23)$$

Remarks Before we give the proof we point out that the triple (b, a, v) is called the *characteristics* of the convolution semigroup, and any Borel measure that satisfies (3.4.23) is called a *Lévy measure* on $\mathbb{R}^d$. Since the proof is quite complex and technical, we will only sketch the main ideas and direct the reader to the literature if they want more details. In particular, we will skate over the measure theoretic technicalities, and try to focus on seeing where b, a and v come from. There are many proofs of this important theorem, and we will use the approach of Zabcyzk [102] pp. 50–55 for the necessity part, as it employs semigroup techniques explicitly. We will not prove the easier sufficiency here, as it requires more probabilistic ideas. Instead we direct the reader to Applebaum [6], p. 30.

Proof. (Necessity) Let $(\mu_t, t \geq 0)$ be a convolution semigroup. Then by Theorem 3.3.10, $(T_t, t \geq 0)$ as given by (3.3.15) is a contraction semigroup on $C_0(\mathbb{R}^d)$. By Theorem 2.2.4 and its proof, for all $\lambda > 0$ the resolvent R_λ exists, and if $A_\lambda = \lambda(\lambda R_\lambda - I)$ is the corresponding Yosida approximant, we have

$$T_t f = \lim_{\lambda \to \infty} e^{t A_\lambda}, \tag{3.4.24}$$

for all $t \geq 0$, $f \in C_0(\mathbb{R}^d)$. Using (1.5.13), we have for all $x \in \mathbb{R}^d$,

$$\lambda R_\lambda f(x) = \lambda \int_0^\infty e^{-\lambda t} \int_{\mathbb{R}^d} f(x+y)\mu_t(dy)dt = \int_{\mathbb{R}^d} f(x+y)\pi_\lambda(dy), \tag{3.4.25}$$

where $\pi_\lambda(A) := \lambda \int_0^\infty e^{-\lambda t} \mu_t(A)dt$, for all $A \in \mathcal{B}(\mathbb{R}^d)$ defines a probability measure on $\mathbb{R}^d$, and we pass over the details of the proof of measurability of the mapping $t \to \mu_t(A)$, which ensures that the last integral is well defined.

Note that for each $\lambda > 0$, $e^{t A_\lambda} f = e^{-\lambda t} e^{t\lambda^2 R_\lambda} f$. The equation (3.4.24) remains valid when we take $f(x) = e^{ix \cdot y}$, for $y \in \mathbb{R}^d$, and when combined with (3.4.25) this yields

$$\Phi_t(y) = \lim_{\lambda \to \infty} \exp\left\{ t \int_{\mathbb{R}^d} (e^{i(x \cdot y)} - 1)\lambda \pi_\lambda(dx) \right\}, \tag{3.4.26}$$

so each term on the right-hand side corresponds to a compound Poisson semigroup with $c = \lambda$ and $\rho = \pi_\lambda$. Define $\eta_\lambda : \mathbb{R}^d \to \mathbb{C}$ by

$$\eta_\lambda(y) := \int_{\mathbb{R}^d} (e^{i(x \cdot y)} - 1)\lambda \pi_\lambda(dx).$$

We now show that η_λ converges pointwise as $\lambda \to \infty$ to a continuous function η. First note that by general properties of characteristic functions, which are easy to prove, we have $\lim_{t \to 0} \Phi_t(y) = 1$ (uniformly for y in bounded sets). So for any $r > 0$ there exists $t > 0$ so that if $|y| < r$, then

$$|\Phi_t(y) - 1| < \frac{1}{4}.$$

We also have that the convergence in (3.4.26) is uniform on finite intervals so that (with t, r as above), there exists $\lambda_0 > 0$ so that for all $\lambda > \lambda_0$,

$$|\Phi_t(y) - e^{t\eta_\lambda}| < \frac{1}{4},$$

and so, combining the last two inequalities, we have (for t and λ as given)

$$|e^{t\eta_\lambda} - 1| < \frac{1}{2}.$$

Hence we can assert that, as $\lambda \to \infty$

$$\Re(\eta_\lambda(y)) = \log_e |e^{\eta_\lambda}| \to \log_e(\Phi_1(y)) = \Re(\eta(y)),$$
$$\Im(\eta_\lambda(y)) = \arg(e^{\eta_\lambda}) \to \arg(\Phi_1(y)) = \Im(\eta(y)),$$

and we thus obtain a well-defined continuous function η, as the last inequality has ensured that we only consider $\arg(z)$ for $z \in \mathbb{C}$ with $|z - 1| < 1/2$.

We must show that η is of the form (3.4.22). Some smoothing is necessary as the finite measure $\nu_\lambda := \lambda \pi_\lambda$ has total mass λ, which diverges as $\lambda \to \infty$. We can try to smooth out this measure by removing some of the mass that accumulates at zero. We do this by replacing ν_λ with the measure $\widetilde{\nu}_\lambda(y) := G(y)\nu_\lambda$, where for all $y \in \mathbb{R}^d$,

$$G(y) := \frac{1}{(2h)^d} \int_{[-h,h]^d} (1 - \cos(u \cdot y)) du$$

$$= \frac{1}{(2h)^d} \int_{[-h,h]^d} (1 - e^{iu \cdot y}) du$$

$$= 1 - \prod_{i=1}^d \frac{\sin(hy_i)}{hy_i} > 0,$$

where $h > 0$ is fixed. Next we show that $\widetilde{\nu}_\lambda$ converges weakly to a finite measure $\widetilde{\nu}$, as $\lambda \to \infty$. We compute the characteristic function, for $y \in \mathbb{R}^d$,

$$\Phi_{\widetilde{\nu}_\lambda}(y) = \int_{\mathbb{R}^d} e^{ix \cdot y} G(x) \nu_\lambda(dx)$$

$$= \int_{\mathbb{R}^d} e^{ix \cdot y} \left(\frac{1}{(2h)^d} \int_{[-h,h]^d} (1 - e^{iu \cdot x}) du \right) \nu_\lambda(dx)$$

$$= \frac{1}{(2h)^d} \int_{[-h,h]^d} (e^{t\eta_\lambda(y+u)} - e^{t\eta_\lambda(y)}) du,$$

where we have used Fubini's theorem. Next we check (using dominated convergence) that we can legitimately take limits to obtain

$$\Psi(y) := \lim_{\lambda \to \infty} \Phi_{\widetilde{\nu}_\lambda}(y) = \frac{1}{(2h)^d} \int_{[-h,h]^d} (e^{t\eta(y+u)} - e^{t\eta(y)}) du.$$

Again using dominated convergence, we can easily check that Ψ is continuous at zero. Then by the Lévy continuity theorem (see, e.g., Dudley [30], pp. 255–8), we can assert that $\widetilde{\nu}_\lambda$ converges weakly to a finite measure $\widetilde{\nu}$, and $\Psi = \Phi_\nu$.

Define $v(dx) := G(x)^{-1}\widetilde{v}(dx)$, then this is our required Lévy measure. To see this observe that since $\sup_{|x|\geq 1} G(x)^{-1} = 1$, we have

$$v(B_1(0)^c) = \int_{B_1(0)^c} G(x)^{-1}\widetilde{v}(dx) < \infty,$$

and since $\lim_{|x|\to 0} \frac{|x|^2}{G(x)} = \frac{h^2}{3}$, it follows that

$$\int_{B_1(0)} |x|^2 v(dx) = \int_{B_1(0)} \frac{|x|^2}{G(x)}\widetilde{v}(dx) < \infty.$$

To extract the terms involving m and a, we introduce cut-offs $0 < r < R < \infty$. Eventually we want to let $R \to \infty$ and $r \to 0$. For each $y \in \mathbb{R}^d$, $\lambda > 0$, we have

$$\eta_\lambda(y) = \int_{|x|<R} (e^{i(x\cdot y)} - 1)v_\lambda(dx) + \int_{|x|\geq R} (e^{i(x\cdot y)} - 1)v_\lambda(dx)$$

$$= \int_{|x|\geq R} (e^{i(x\cdot y)} - 1)v_\lambda(dx) + i\int_{|x|<R} \sin(x\cdot y)v_\lambda(dx)$$

$$+ \int_{|x|<r} (\cos(x\cdot y) - 1)v_\lambda(dx)$$

$$+ \int_{r\leq|x|<R} (\cos(x\cdot y) - 1)v_\lambda(dx).$$

There are no problems in taking limits as $\lambda \to \infty$ in the first term to obtain

$$\lim_{\lambda\to\infty} \int_{|x|\geq R} (e^{i(x\cdot y)} - 1)v_\lambda(dx) = \int_{|x|\geq R} (e^{i(x\cdot y)} - 1)v(dx), \qquad (3.4.27)$$

and this term will go to zero when we, eventually, take the limit $R \to \infty$.

For the final term, as is easily verified, there exists $C > 0$ so that

$$|1 - \cos(x\cdot y)| \leq C(|x|^2 \wedge 1).$$

So the given function is v-integrable, and we have

$$\lim_{r\to 0} \lim_{\lambda\to\infty} \int_{r\leq|x|<R} (\cos(x\cdot y) - 1)v_\lambda(dx) = \int_{|x|<R} (\cos(x\cdot y) - 1)v(dx).$$

$$(3.4.28)$$

The other two terms are more complicated. We write

$$\int_{|x|<R} \sin(x\cdot y)v_\lambda(dx) = I_1(\lambda)(y) + I_2(\lambda)(y),$$

where

$$I_1(\lambda)(y) := \int_{|x|<R} \frac{\sin(x\cdot y) - (x\cdot y)/(1 + |x|^2)}{|x|^2} \cdot |x|^2 G^{-1}(x)\widetilde{v}_\lambda(dx)$$

and

$$I_2(\lambda)(y) := y \cdot \int_{|x|<R} x/(1+|x|^2)\nu_\lambda(dx).$$

Now $I_1(\lambda)(y)$ is of the form $\int_{|x|<R} p_y(x)q(x)\tilde{\nu}(dx)$, where $p_y, q \in C_b(\mathbb{R}^d)$ and hence the integral converges to give

$$\lim_{\lambda\to\infty} I_1(\lambda)(y) := \int_{|x|<R} \left(\sin(x\cdot y) - \frac{x\cdot y}{1+|y|^2}\right)\nu(dy). \qquad (3.4.29)$$

But then $I_2(\lambda)(y)$ must converge as $\lambda \to \infty$, and we deduce that there exists $m \in \mathbb{R}^d$ so that

$$\lim_{\lambda\to\infty} I_2(\lambda)(y) = m \cdot y. \qquad (3.4.30)$$

Similarly we write

$$\int_{|x|<r} (\cos(x\cdot y) - 1))\nu_\lambda(dx) = I_3(\lambda)(y) + I_4(\lambda)(y),$$

where

$$I_3(\lambda)(y) = \int_{|x|<r} \frac{\cos(x\cdot y) - 1 + (x\cdot y)^2/2}{|y|^2}|y|^2 G(y)^{-1}\tilde{\nu}_\lambda(dx)$$

and

$$I_4(\lambda)(y) := -\frac{1}{2}\int_{|x|<r}(x\cdot y)^2\nu_\lambda(dx).$$

As before we have

$$\lim_{\lambda\to\infty} I_3(\lambda)(y) = \int_{|y|<r}(\cos(x\cdot y) - 1 + (x\cdot y)^2/2)\nu(dy),$$

and this term will tend to zero when we take the limit as $r \to 0$. We hence find that

$$\lim_{\lambda\to\infty} I_3(\lambda)(y) = -\frac{1}{2}\int_{|x|<r}(x\cdot y)^2\nu(dx).$$

If we define a matrix $a(r)$ by $a_{ij}(r) = \frac{1}{2}\int_{|x|<r}(x\cdot e_j)(x\cdot e_i)\nu(dx)$, then we have

$$a(r)y\cdot y = \frac{1}{2}\int_{|x|<r}(x\cdot y)^2\nu(dx),$$

and then our required matrix a is determined by $ay\cdot y = \lim_{r\to 0} a(r)y\cdot y$.

Finally we combine this last limit with (3.4.27), (3.4.28) and (3.4.29), and let $R \to \infty$ to get the required result. $\square$

The formula (3.4.22) is called the *Lévy–Khintchine formula*. In the sequel, it will be convenient to write it in a slightly different way. The mapping $x \to \left(\frac{1}{1+|x|^2} - \mathbf{1}_{B_1}(x)\right) x$ is easily seen to be ν-integrable; thus we may rewrite (3.4.22) as

$$\eta(y) = ib \cdot y - ay \cdot y$$
$$+ \int_{\mathbb{R}^d} (e^{ix \cdot y} - 1 - ix \cdot y \mathbf{1}_{B_1}(x))\nu(dx), \qquad (3.4.31)$$

where $b := m - \int_{\mathbb{R}^d} \left(\frac{1}{1+|x|^2} - \mathbf{1}_{B_1}(x)\right) x\nu(dx)$. The mapping $\eta \colon \mathbb{R}^d \to \mathbb{C}$ is called the *characteristic exponent* or *Lévy symbol* of the convolution semigroup $(\mu_t, t \geq 0)$. We will find a use later for the following estimate (which you can derive in Problem 3.5): there exists $K > 0$, so that for all $y \in \mathbb{R}^d$,

$$|\eta(y)| \leq K(1 + |y|^2). \qquad (3.4.32)$$

To derive this, it helps to write the final integral in (3.4.31) as

$$\int_{||x||<1} (e^{ix \cdot y} - 1 - ix \cdot y)\nu(dx) + \int_{||x|| \geq 1} (e^{ix \cdot y} - 1)\nu(dx),$$

and use well-known estimates such as

$$|e^{ix \cdot y} - 1 - ix \cdot y| \leq C|x|^2,$$

where $C > 0$, for all $x, y \in \mathbb{R}^d$.

3.4.1 Stable Semigroups

As an important application of the Lévy–Khintchine formula, we consider convolution semigroups indexed by a parameter $0 < \alpha < 2$ for which $a = 0$ and there exists $C_1, C_2 \geq 0$ with $C_1 + C_2 > 0$ and $\nu(dx) = \nu(x)dx$ where

$$\nu(x) = \frac{C_1}{|x|^{d+\alpha}} \mathbf{1}_{\{|x|<0\}}(x) + \frac{C_2}{|x|^{d+\alpha}} \mathbf{1}_{\{|x| \geq 0\}}(x). \qquad (3.4.33)$$

The corresponding convolution semigroups are called the α-stable laws, and they have many applications in both theoretical and applied probability. They are a parameterised family of semigroups that include both the Cauchy distribution ($\alpha = 1$) and the Gauss–Weierstrass function ($\alpha = 2$).

For simplicity, we will choose the case $\alpha < 1$, $C_1 = 0$ and $d = 1$ so we deal with the so-called *one-sided α-stable subordinator*.[9] In this case, we easily

---

[9] The *subordination* idea leads to a lot of very interesting mathematics that we do not have space to pursue here. For a brief introduction, see subsection 1.3.2 of Applebaum [6], pp. 52–62. For a monograph treatment, see Schilling et al. [88].

check that $\int_0^1 xv(dx) < \infty$, and so we may take $b = -\int_0^1 ydy$, to find that η, as given by (3.4.31), takes the form

$$\eta(y) = C \int_{(0,\infty)} \frac{e^{ixy} - 1}{x^{1+\alpha}} dx, \tag{3.4.34}$$

where $C := C_2$.

To simplify (3.4.34), we introduce the density $f^{\alpha,\beta}$ for the well-known gamma distribution on $(0, \infty)$ with parameters $a, b > 0$:

$$f^{a,b}(x) = \frac{b^a}{\Gamma(a)} x^{a-1} e^{-bx},$$

for $x > 0$. It is an easy calculation to show that its characteristic function $\Phi^{a,b}$ takes the following form for all $y \in \mathbb{R}$:

$$\Phi^{a,b}(y) = \left(\frac{b}{b - iy} \right)^a. \tag{3.4.35}$$

Now the trick to simplify (3.4.34) is to introduce an extra parameter $s > 0$ which will eventually be sent to zero (see Meerschaert and Sikorskii [66] pp. 56–7). We define

$$\eta_s(y) = C_\alpha \int_{(0,\infty)} \frac{e^{(iy-s)x} - 1}{x^{1+\alpha}} dx,$$

where $C_\alpha := C/\Gamma(1 - \alpha)$. Integration by parts and then use of (3.4.35) yields

$$\eta_s(y) = C_\alpha (iy - s) \int_{(0,\infty)} e^{ixy} e^{-sx} x^{-\alpha} dx$$

$$= C_\alpha (iy - s) \frac{\Gamma(1 - \alpha)}{s^\alpha} \int_{(0,\infty)} e^{ixy} f^{1-\alpha,s}(x) dx$$

$$= -C(s - iy)^\alpha.$$

Now, by use of dominated convergence, we conclude that

$$\eta(y) = \lim_{s \to 0} \eta_s(y) = -C(-iy)^\alpha. \tag{3.4.36}$$

In this case we can also make the analytic continuation $y \to iy$ to compute the Laplace transform (for $y > 0$)

$$\int_{(0,\infty)} e^{-xy} \mu_t(dx) = e^{-tCy^\alpha}, \tag{3.4.37}$$

and this will be useful later.

More generally, and returning to the general case of $d \geq 1$, we say that a Borel measure λ on $\mathbb{R}^d$ is *rotationally invariant* if $\lambda(OA) = \lambda(A)$ for all $A \in \mathcal{B}(\mathbb{R}^d)$ and $O \in O(d)$, where $O(d)$ is the group of $d \times d$ real orthogonal matrices, i.e., $O \in O(d)$ if and only if $O^T O = I$. In the case $d = 1$, a measure is rotationally invariant if and only if it is *symmetric*, i.e., $\lambda(-A) = \lambda(A)$ for all $A \in \mathcal{B}(\mathbb{R}^d)$, where $-A := \{-x, x \in A\}$. You can easily check that a probability measure is rotationally invariant if and only if its characteristic function is, i.e., $\Phi_\lambda(Oy) = \Phi_\lambda(y)$, for all $O \in O(d)$, $y \in \mathbb{R}^d$. There has been a great deal of interest in recent years in the C_0-semigroups that arise from rotationally invariant α-stable convolution semigroups where $0 < \alpha \leq 2$. It can be shown that η is the characteristic exponent of such a semigroup if and only if, for all $y \in \mathbb{R}^d$,

$$\eta(y) = -k|y|^\alpha, \qquad (3.4.38)$$

where $k > 0$. A proof can be found on p. 86 of Sato [84]. the discerning reader will note that I have now included the case $\alpha = 2$, whereas I did not before. There are good probabilistic reasons for doing this, but we will not pursue them here. If we put $\alpha = 2$ in (3.4.38) then we have a convolution semigroup with characteristics $(0, kI, 0)$, and this is the signature of a rescaled Brownian motion. Note also that $\alpha = 1$ corresponds to a rescaled Cauchy process.

It is worth pointing out that the stable convolution semigroups all have densities for $t > 0$. In the case $d = 1$, series expansions for these are given in Feller [35], pp. 548–50.

3.4.2 Lévy Processes

We have seen that the Gauss–Weierstrass and Cauchy functions give rise to convolution semigroups, and each of these are associated to a process. More generally, given any convolution semigroup $(\mu_t, t \geq 0)$, there exists a process $L = (L(t), t \geq 0)$ such that $P(L(t) \in A) = \mu_t(A)$ for all $A \in \mathcal{B}(\mathbb{R}^d)$. These processes are called *Lévy processes* and they are defined as follows:

(L1) The process L has stationary and independent increments, i.e., the random variables $L(t_1) - L(t_0), L(t_2) - L(t_1), \ldots, L(t_n) - L(t_{n-1})$ are independent for all $0 = t_0 < t_1 < \cdots < t_n, n \in \mathbb{N}$, and $P(L(t) - L(s) \in A) = P(L(t - s) - L(0) \in A)$ for all $0 < s \leq t < \infty, A \in \mathcal{B}(\mathbb{R}^d)$.

(L2) $L(0) = 0$ (a.s.).

(L3) The process is *stochastically continuous*, i.e., $\lim_{t \to 0} P(|L(t)| > a) = 0$ for all $a > 0$.

(L4) There exists $\Omega' \in \mathcal{F}$ with $P(\Omega') = 1$ so that the mapping $t \to L(t)\omega$ is right continuous with left limits from $[0, \infty)$ to $\mathbb{R}^d$ for all $\omega \in \Omega'$.

For a very nice introduction to these processes, see Schilling [86], while detailed monograph accounts may be found in Kyprianou [61] and Sato [84].

It can be shown that (L4) can be sharpened to sample path continuity (as in (B4)) if and only if L is a Brownian motion with drift, and so has characteristics $(b, a, 0)$. In general the paths of a generic Lévy process look like a Brownian motion with drift but with jump discontinuities of arbitrary random size occurring at random times. The intensity of jumps is described by the Lévy measure ν. The compound Poisson semigroup is associated to the *compound Poisson process* $Y = (Y(t), t \geq 0)$ which is described as follows. We require a sequence $(X_n, n \in \mathbb{N})$ of independent, identically distributed random variables having common law ρ and an independent Poisson process $N = (N(t), n \in \mathbb{N})$ with intensity c, so N takes values in $\mathbb{Z}_+$ with $P(N(t) = n) = e^{-ct} \frac{(ct)^n}{n!}$. We then define

$$Y(t) = \begin{cases} X_1 + X_2 + \cdots + X_{N(t)} & \text{if } t > 0, \\ 0 & \text{if } t = 0. \end{cases}$$

In this case the Lévy measure $\nu = c\rho$ and the process has characteristics $(c \int_0^1 x\rho(dx), 0, c\rho)$. We may also introduce a Brownian motion with drift that is independent of the compound Poisson process and so we have a Lévy process taking the form $L(t) = mt + \sigma B(t) + Y(t)$, for $t \geq 0$. This looks more like a generic Lévy process. Its characteristic function is given by (3.3.21) and we see that the times of jumps are those of the Poisson process N and the size of the nth jump is the value of the random variable X_n. The stable Lévy processes have the useful property of being *self-similar* with Hurst index $1/\alpha$, i.e., for each $t \geq 0, c > 0$, the random variable X_{ct} has the same distribution as $c^{1/\alpha} X_t$.

We now give a formal proof that the laws of a Lévy process really do yield a convolution semigroup.

Proposition 3.4.13 *If $L = (L(t), t \geq 0)$ is a Lévy process and μ_t is the law of $L(t)$, then $(\mu_t, t \geq 0)$ is a convolution semigroup.*

Proof. We have $\mu_0 = \delta_0$ by (L2). Using (L1) we have for all $f \in B_b(\mathbb{R}^d), s, t \geq 0$,

$$\int_{\mathbb{R}^d} f(x)\mu_{s+t}(dx) = \mathbb{E}(f(L(s + t)))$$
$$= \mathbb{E}(f(L(t + s) - L(s) + L(s)))$$
$$= \mathbb{E}(f(L(s + t) - L(s) + y | L(s) = y)))\mu_s(dy)$$
$$= \int_{\mathbb{R}^d} \int_{\mathbb{R}^d} f(y + z)\mu_t(dz)\mu_s(dy),$$

and so $\mu_{s+t} = \mu_s * \mu_t$. Finally to show weak continuity, suppose that $f \in C_b(\mathbb{R}^d)$ with $f \neq 0$; then given any $\epsilon > 0$ there exists $\delta > 0$ such that

$$\sup_{x \in B_\delta(0)} |f(x) - f(0)| < \epsilon/2$$

and by (L3) there exists $\delta' > 0$ such that

$$0 < t < \delta' \Rightarrow P(|X(t)| > \delta) < \frac{\epsilon}{4M},$$

where $M := \sup_{x \in \mathbb{R}^d} |f(x)|$. For such t we then find that

$$\left| \int_{\mathbb{R}^d} (f(x) - f(0)) \mu_t(dx) \right|$$

$$\leq \int_{B_\delta(0)} |f(x) - f(0)| \mu_t(dx) + \int_{B_\delta(0)^c} |f(x) - f(0)| \mu_t(dx)$$

$$\leq \sup_{x \in B_\delta(0)} |f(x) - f(0)| + 2M P(X(t) \in B_\delta(0)^c)$$

$$< \epsilon. \qquad \qquad \square$$

The converse statement, that given any convolution semigroup $(\mu_t, t \geq 0)$ there exists a Lévy process L so that μ_t is the law of $L(t)$ is true but will not be pursued here (see, e.g., Theorem 7.10 in Sato [84] pp. 35–7).

Now just as in the case of the Brownian and Cauchy processes, we can write the semigroup $(T_t, t \geq 0)$ of (3.3.15) in the form

$$T_t f(x) = (f * \mu_t)(x) = \mathbb{E}(f(L(t) + x)).$$

Next we will find the generators of these semigroups.

3.5 Generators of Convolution Semigroups

This section is very closely based on material in Applebaum [6], pp. 163–76.

We define *Schwartz space* $S(\mathbb{R}^d)$ to be the linear space of all $f \in C^\infty(\mathbb{R}^d)$ for which

$$\sup_{x \in \mathbb{R}^d} |x^\beta D^\alpha f(x)| < \infty$$

for all multi-indices α and β. Note that "Gaussian functions" of the form $x \to e^{-c|x|^2}$ for $c > 0$ are in $S(\mathbb{R}^d)$. It is a fact that $S(\mathbb{R}^d) \subseteq C_0(\mathbb{R}^d)$ and also $S(\mathbb{R}^d) \subseteq L^p(\mathbb{R}^d, \mathbb{R})$ for all $1 \leq p < \infty$ (see Problem 3.7). Since $C_c^\infty(\mathbb{R}^d) \subseteq S(\mathbb{R}^d)$, it follows that Schwartz space is dense in each of these Banach spaces. Similar comments apply to $S(\mathbb{R}^d, \mathbb{C})$, which is defined analogously.

One of the reasons for working with Schwartz space in the sequel, rather than, say $C_c(\mathbb{R}^d)$, is that the Fourier transform acts on $S(\mathbb{R}^d, \mathbb{C})$ as a bijection (which is even continuous, when the Schwartz space is equipped with a suitable topology).

Let $f \in S(\mathbb{R}^d)$. We recall that its Fourier transform is $\hat{f} \in S(\mathbb{R}^d, \mathbb{C})$, where

$$\hat{f}(u) = (2\pi)^{-d/2} \int_{\mathbb{R}^d} e^{-i(u,x)} f(x) dx$$

for all $u \in \mathbb{R}^d$, and the Fourier inversion formula yields

$$f(x) = (2\pi)^{-d/2} \int_{\mathbb{R}^d} \hat{f}(u) e^{i(u,x)} du$$

for each $x \in \mathbb{R}^d$ (see Appendix B for a swift proof).

The next theorem uses the language of pseudo-differential operators. For those who have not met these before, the following subsection provides a motivational introduction.

Theorem 3.5.14 *Let $(\mu_t, t \geq 0)$ be a convolution semigroup with characteristic exponent η and characteristics (b, a, v). Let $(T_t, t \geq 0)$ be the associated C_0-semigroup and A be its infinitesimal generator.*

1. For each $t \geq 0$, $f \in S(\mathbb{R}^d)$, $x \in \mathbb{R}^d$,

$$(T_t f)(x) = (2\pi)^{-d/2} \int_{\mathbb{R}^d} e^{iu \cdot x} e^{t\eta(u)} \hat{f}(u) du,$$

so that T_t is a pseudo-differential operator with symbol $e^{t\eta}$.
2. For each $f \in S(\mathbb{R}^d)$, $x \in \mathbb{R}^d$,

$$(Af)(x) = (2\pi)^{-d/2} \int_{\mathbb{R}^d} e^{iu \cdot x} \eta(u) \hat{f}(u) du,$$

so that A is a pseudo-differential operator with symbol η.
3. For each $f \in S(\mathbb{R}^d)$, $x \in \mathbb{R}^d$,

$$(Af)(x) = \sum_{i=1}^{d} b_i \partial_i f(x) + \sum_{i,j=1}^{d} a_{ij} \partial_i \partial_j f(x)$$

$$+ \int_{\mathbb{R}^d - \{0\}} [f(x+y) - f(x) - \sum_{i=1}^{d} y_i \partial_i f(x) \mathbf{1}_{B_1}(y)] v(dy).$$

$$(3.5.39)$$

Proof. (1) We apply Fourier inversion within (3.3.15) to find that for all $t \geq 0$, $f \in S(\mathbb{R}^d)$, $x \in \mathbb{R}^d$,

$$(T_t f)(x) = \mathbb{E}\big(f(X(t) + x)\big) = (2\pi)^{-d/2} \mathbb{E}\left(\int_{\mathbb{R}^d} e^{i(u, x + X(t))} \hat{f}(u) du\right).$$

Since $\hat{f} \in S(\mathbb{R}^d)$ to $S)(\mathbb{R}^d, \mathbb{C})$ and change $L^p(\mathbb{R}^d)$ to $L^p(\mathbb{R}^d, \mathbb{C})$. we have

$$\left| \int_{\mathbb{R}^d} e^{iu \cdot x} \mathbb{E}(e^{i(u, X(t))}) \, \hat{f}(u) du \right| \leq \int_{\mathbb{R}^d} |e^{iu \cdot x} \mathbb{E}(e^{i(u, X(t))})| |\hat{f}(u)| du$$

$$\leq \int_{\mathbb{R}^d} |\hat{f}(u)| du < \infty,$$

so we can apply Fubini's theorem to obtain

$$(T_t f)(x) = (2\pi)^{-d/2} \int_{\mathbb{R}^d} e^{iu \cdot x} \mathbb{E}(e^{i(u, X(t))}) \hat{f}(u) du$$

$$= (2\pi)^{-d/2} \int_{\mathbb{R}^d} e^{iu \cdot x} e^{t\eta(u)} \hat{f}(u) du.$$

(2) For each $f \in S(\mathbb{R}^d)$, $x \in \mathbb{R}^d$, we have by result (1),

$$(Af)(x) = \lim_{t \to 0} \frac{1}{t} [(T_t f)(x) - f(x)]$$

$$= (2\pi)^{-d/2} \lim_{t \to 0} \int_{\mathbb{R}^d} e^{iu \cdot x} \frac{e^{t\eta(u)} - 1}{t} \hat{f}(u) du.$$

Now, by (3.4.32), there exists $K > 0$ such that

$$\int_{\mathbb{R}^d} \left| e^{iu \cdot x} \frac{e^{t\eta(u)} - 1}{t} \hat{f}(u) \right| du \leq \int_{\mathbb{R}^d} |\eta(u) \hat{f}(u)| du$$

$$\leq K \int_{\mathbb{R}^d} (1 + |u|^2|) |\hat{f}(u)| du < \infty,$$

since $(1 + |u|^2) \hat{f}(u) \in S(\mathbb{R}^d, \mathbb{C})$.

We can now use dominated convergence to deduce the required result.

(3) Applying the Lévy–Khintchine formula to result (2), we obtain for each $f \in S(\mathbb{R}^d)$, $x \in \mathbb{R}^d$,

$$(Af)(x) = (2\pi)^{-d/2} \int_{\mathbb{R}^d} e^{ix \cdot u} \left\{ ib \cdot u - au \cdot u \right.$$

$$\left. + \int_{\mathbb{R}^d - \{0\}} [e^{iu \cdot y} - 1 - iu \cdot y \mathbf{1}_{B_1(0)}(y)] v(dy) \right\} \hat{f}(u) du.$$

The required result now follows immediately from elementary properties of the Fourier transform. An interchange of integrals is required, but this is justified by Fubini's theorem in a similar way to the arguments given above. □

The alert reader will have noticed that we appear to have cheated in our proof of (2), in that we have computed the generator using the pointwise limit instead of the uniform one. In fact the operators defined by both limits coincide in this context; see Sato [84], Lemma 31.7, p. 209. A linear operator in $C_0(\mathbb{R}^d)$ that has the form (3.5.39) will be called a *Lévy operator*.

Examples 1. Fix $b \in \mathbb{R}^d$ and consider the convolution semigroup $\{\delta_{bt}, t \geq 0\}$. This gives rise to a slight generalisation of the translation semigroup of Example 1.2.5:

$$T_t f(x) = f(x + bt),$$

for all $f \in C_0(\mathbb{R}^d), x \in \mathbb{R}^d, t \geq 0$. The characteristic function is easily obtained to give $\Phi_{\delta_{bt}}(y) = e^{itb \cdot y}$, and so the characteristics are $(b, 0, 0)$. Hence by Theorem 3.5.14, the action of the generator A on $S(\mathbb{R}^d)$ is:

$$A = b \cdot \nabla = \sum_{i=1}^{d} b_i \partial_i.$$

This could also have been found directly by using a Taylor series expansion.

2. A general diffusion operator (with constant coefficients) is associated to a Gaussian convolution semigroup having characteristics $(b, a, 0)$. In this case, the action of the generator on $S(\mathbb{R}^d)$ is given by

$$A = \sum_{i=1}^{d} b_i \partial_i + \sum_{i,j=1}^{d} a_{ij} \partial_i \partial_j.$$

In particular, the choice of characteristics $(0, I, 0)$ returns us to the world of heat kernels and Laplacians with which we started the chapter.

3. If we consider the example of the rotationally invariant α-stable semigroup, where $0 < \alpha < 2$, we have $\eta(u) = -|u|^{\alpha}$. But then we may interpret $A = \eta(D) = -(-\Delta)^{\alpha/2}$. If we take $\alpha = 1$ to return to the case of the Cauchy distribution, then we have justified the notation that we used in subsection 3.1.3.

3.5.1 Lévy Generators as Pseudo-Differential Operators

We have freely used the terminology "pseudo-differential operators" when discussing Lévy generators in the last section. What do we mean by this?

We begin by examining the Fourier transform of differential operators. More or less everything flows from the following simple fact:

$$D^{\alpha} e^{iu \cdot x} = u^{\alpha} e^{iu \cdot x},$$

for each $x, u \in \mathbb{R}^d$ and each multi-index α.

Using Fourier inversion and dominated convergence, we then find that

$$(D^\alpha f)(x) = (2\pi)^{-d/2} \int_{\mathbb{R}^d} u^\alpha \hat{f}(u) e^{iu \cdot x} du$$

for all $f \in S(\mathbb{R}^d, \mathbb{C})$, $x \in \mathbb{R}^d$.

If p is a polynomial in u of the form $p(u) = \sum_{|\alpha| \le k} c_\alpha u^\alpha$, where $k \in \mathbb{N}$ and each $c_\alpha \in \mathbb{C}$, we can form the associated differential operator $P(D) = \sum_{|\alpha| \le k} c_\alpha D^\alpha$ and, by linearity,

$$(P(D)f)(x) = (2\pi)^{-d/2} \int_{\mathbb{R}^d} p(u) \hat{f}(u) e^{iu \cdot x} du.$$

The next step is to employ variable coefficients. If each $c_\alpha \in C^\infty(\mathbb{R}^d)$, for example, we may define $p(x, u) = \sum_{|\alpha| \le k} c_\alpha(x) u^\alpha$ and $P(x, D) = \sum_{|\alpha| \le k} c_\alpha(x) D^\alpha$. We then find that

$$(P(x, D)f)(x) = (2\pi)^{-d/2} \int_{\mathbb{R}^d} p(x, u) \hat{f}(u) e^{iu \cdot x} du.$$

The passage from D to $P(x, D)$ has been rather straightforward, but now we will take a leap into the unknown and abandon formal notions of differentiation. So we replace p by a more general function $\sigma \colon \mathbb{R}^d \times \mathbb{R}^d \to \mathbb{C}$. Informally, we may then define a *pseudo-differential operator* $\sigma(x, D)$ by the prescription:

$$(\sigma(x, D)f)(x) = (2\pi)^{-d/2} \int_{\mathbb{R}^d} \sigma(x, u) \hat{f}(u) e^{iu \cdot x} du,$$

and σ is then called the *symbol* of this operator. We have been somewhat cavalier here, and we should make some further assumptions on the symbol σ to ensure that $\sigma(x, D)$ really is a bona fide operator. There are various classes of symbols that may be defined to achieve this, but we will not pursue that theme here. Instead we refer the reader to standard references, such as Chapter 2 of Ruzhansky and Turunen [83] or Taylor [97].

Much of the motivation for introducing pseudo-differential operators came from partial differential equations. Let us look informally at a motivating example. Suppose we want to solve the elliptic equation $\Delta f = g$, with $g \in S(\mathbb{R}^d)$. Informally, "$f = \Delta^{-1} g$", and we would like to give meaning to "Δ^{-1}". If we take Fourier transforms of both sides of the equation we obtain, for each $y \in \mathbb{R}^d$,

$$-|y|^2 \hat{f}(y) = \widehat{\Delta f}(y) = \hat{g}(y),$$

and so $\widehat{f}(y) = -1/|y|^2 \widehat{g}(y)$ (for $y \neq 0$). Now apply the Fourier inversion formula to obtain, for each $x \in \mathbb{R}^d$

$$f(x) = (2\pi)^{-d/2} \int_{\mathbb{R}^d} \sigma(y)\widehat{g}(y)e^{i(y,x)}du,$$

and the right-hand side is the action of a pseudo-differential operator with symbol $\sigma(y) = -1/|y|^2$.

If we return to the case of convolution semigroups $(\mu_t, t \geq 0)$ having symbol η, we can interpret the generator A as the operator $\eta(D)$, and so the semigroup $(T_t, t \geq 0)$ implements the (unique) solution $u(t) = T_t f$, for $f \in C_0(\mathbb{R}^d)$, of a generalised partial differential equation:

$$\frac{\partial u(t)}{\partial t} = \eta(D)u(t),$$

where $u(0) = f$. If the characteristics of the convolution semigroup are (b, a, ν), then we obtain a second-order (parabolic) partial differential differentiation with constant coefficients if and only if $\nu = 0$.

3.6 Extension to L^p

We conclude this chapter by showing that our semigroups may be extended to $L^p(\mathbb{R}^d)$ for $1 \leq p < \infty$.

Theorem 3.6.15 *If $(\mu_t, t \geq 0)$ is a convolution semigroup in $\mathbb{R}^d$, then for each $1 \leq p < \infty$, the prescription*

$$(T_t f)(x) = \int_{\mathbb{R}^d} f(x+y)\mu_t(dy)$$

$f \in L^p(\mathbb{R}^d)$, $x \in \mathbb{R}^d$, $t \geq 0$ gives rise to a C_0-semigroup $(T_t \geq 0)$.

Proof. We must show that each $T_t : L^p(\mathbb{R}^d) \to L^p(\mathbb{R}^d)$. In fact, for all $f \in L^p(\mathbb{R}^d)$, $t \geq 0$, by Jensen's inequality (or Hölder's inequality if you prefer) and Fubini's theorem, we obtain

$$\|T_t f\|_p^p = \int_{\mathbb{R}^d} \left| \int_{\mathbb{R}^d} f(x+y)q_t(dy) \right|^p dx$$

$$\leq \int_{\mathbb{R}^d} \int_{\mathbb{R}^d} |f(x+y)|^p q_t(dy)dx$$

$$= \int_{\mathbb{R}^d} \left(\int_{\mathbb{R}^d} |f(x+y)|^p dx \right) q_t(dy)$$

$$= \int_{\mathbb{R}^d} \left(\int_{\mathbb{R}^d} |f(x)|^p dx \right) q_t(dy) = \|f\|_p^p,$$

and we have proved that each T_t is a contraction in $L^p(\mathbb{R}^d)$.

Now we need to establish the semigroup property $T_{s+t}f = T_s T_t f$ for all $s, t \geq 0$. We know that this holds for all $f \in C_0(\mathbb{R}^d) \cap L^p(\mathbb{R}^d)$. However, this space is dense in $L^p(\mathbb{R}^d)$, and the result follows by continuity since each T_t is bounded.

Finally, we must prove strong continuity. First we let $f \in C_c(\mathbb{R}^d)$ and choose a ball B centred on the origin in $\mathbb{R}^d$. Then, using Jensen's inequality and Fubini's theorem as above, we obtain for each $t \geq 0$

$$\|T_t f - f\|_p^p = \int_{\mathbb{R}^d} \left| \int_{\mathbb{R}^d} [f(x+y) - f(x)] \mu_t(dy) \right|^p dx$$

$$\leq \int_B \left(\int_{\mathbb{R}^d} |f(x+y) - f(x)|^p dx \right) \mu_t(dy)$$

$$+ \int_{B^c} \left(\int_{\mathbb{R}^d} |f(x+y) - f(x)|^p dx \right) \mu_t(dy)$$

$$\leq \int_B \left(\int_{\mathbb{R}^d} |f(x+y) - f(x)|^p dx \right) \mu_t(dy)$$

$$+ \int_{B^c} \left(\int_{\mathbb{R}^d} 2^p \max\{|f(x+y)|^p, |f(x)|^p\} dx \right) \mu_t(dy)$$

$$\leq \sup_{y \in B} \int_{\mathbb{R}^d} |f(x+y) - f(x)|^p dx + 2^p \|f\|_p^p \mu_t(B^c).$$

By choosing B to have sufficiently small radius, we obtain $\lim_{t \to 0} \|T_t f - f\|_p = 0$ from the continuity of f and dominated convergence in the first term and the weak continuity of $(\mu_t, t \geq 0)$ in the second term, just as in the proof of Theorem 3.3.10.

Now let $f \in L^p(\mathbb{R}^d)$ be arbitrary and choose a sequence $(f_n, n \in \mathbb{N})$ in $C_c(\mathbb{R}^d)$ that converges to f. Using the triangle inequality and the fact that each T_t is a contraction, we obtain, for each $t \geq 0$,

$$\|T_t f - f\| \leq \|T_t f_n - f_n\| + \|T_t(f - f_n)\| + \|f - f_n\|$$

$$\leq \|T_t f_n - f_n\| + 2\|f - f_n\|, \qquad \square$$

from which the required result follows.

For the case $p = 2$ we can explicitly compute the domain of the infinitesimal generator. First, using the fact that the Fourier transform is a unitary isomorphism of $L^2(\mathbb{R}^d; \mathbb{C})$, we show that

$$(T_t f)(x) = (2\pi)^{-d/2} \int_{\mathbb{R}^d} e^{iu \cdot x} e^{t\eta(u)} \hat{f}(u) du \qquad (3.6.40)$$

for all $t \geq 0$, $x \in \mathbb{R}^d$, $f \in L^2(\mathbb{R}^d)$. This is the L^2-counterpart of Theorem 3.5.14 (1).

Define $\mathcal{H}_\eta(\mathbb{R}^d) = \left\{ f \in L^2(\mathbb{R}^d); \int_{\mathbb{R}^d} |\eta(u)|^2 |\hat{f}(u)|^2 du < \infty \right\}$. Then we have

Theorem 3.6.16 $D_A = \mathcal{H}_\eta(\mathbb{R}^d)$.

Proof. We follow Berg and Forst [11], p. 92. Let $f \in D_A$; then $Af = \lim_{t\downarrow 0}[(1/t)(T_t f - f)]$ in $L^2(\mathbb{R}^d)$. We take Fourier transforms and use the continuity of $\mathcal{F}$ to obtain

$$\widehat{Af} = \lim_{t\to 0} \frac{1}{t}(\widehat{T_t f} - \hat{f}).$$

Using (3.6.40), we have

$$\widehat{Af} = \lim_{t\to 0} \frac{1}{t}(e^{t\eta}\hat{f} - \hat{f});$$

hence, for any sequence $(t_n, n \in \mathbb{N})$ in $\mathbb{R}^+$ for which $\lim_{n\to\infty} t_n = 0$, we get

$$\widehat{Af} = \lim_{n\to\infty} \frac{1}{t_n}(e^{t_n\eta}\hat{f} - \hat{f}). \qquad \text{a.e.}$$

However, $\lim_{n\to\infty}[(1/t_n)(e^{t_n\eta} - 1)] = \eta$ and so $\widehat{Af} = \eta\hat{f}$ (a.e.). But then $\eta\hat{f} \in L^2(\mathbb{R}^d; \mathbb{C})$, i.e., $f \in \mathcal{H}_\eta(\mathbb{R}^d)$.

So we have established that $D_A \subseteq \mathcal{H}_\eta(\mathbb{R}^d)$.

Conversely, let $f \in \mathcal{H}_\eta(\mathbb{R}^d)$; then by (3.6.40) again,

$$\lim_{t\to 0} \frac{1}{t}(\widehat{T_t f} - \hat{f}) = \lim_{t\to 0} \frac{1}{t}(e^{t\eta}\hat{f} - \hat{f}) = \eta\hat{f} \in L^2(\mathbb{R}^d).$$

Hence, by the unitarity and continuity of the Fourier transform, $\lim_{t\to 0}[(1/t)(T_t f - f)] \in L^2(\mathbb{R}^d; \mathbb{C})$ and so $f \in D_A$. $\qquad \square$

Readers should note that the proof has also established the pseudo-differential operator representation

$$Af = (2\pi)^{-d/2} \int_{\mathbb{R}^d} e^{iu\cdot x}\eta(u)\hat{f}(u)du \qquad (3.6.41)$$

for all $f \in \mathcal{H}_\eta(\mathbb{R}^d)$.

The space $\mathcal{H}_\eta(\mathbb{R}^d)$ is called an *anisotropic Sobolev space* by Jacob [53]. It is clearly non-empty, indeed $S(\mathbb{R}^d) \subseteq \mathcal{H}_\eta(\mathbb{R}^d)$. Note that if we take X to be a standard Brownian motion then $\eta(u) = -\frac{1}{2}|u|^2$ for all $u \in \mathbb{R}^d$ and

$$\mathcal{H}_\eta(\mathbb{R}^d) = \left\{ f \in L^2(\mathbb{R}^d); \int_{\mathbb{R}^d} |u|^4 |\hat{f}(u)|^2 du < \infty \right\}.$$

This is precisely the usual Sobolev space, which is usually denoted $\mathcal{H}_2(\mathbb{R}^d)$ and which can be defined equivalently (see Appendix C) as the completion of $C_c^\infty(\mathbb{R}^d)$ with respect to the norm

$$\|f\|_2 = \left(\int_{\mathbb{R}^d} (1 + |u|^2)^2 |\hat{f}(u)|^2 du \right)^{1/2}$$

for each $f \in C_c^\infty(\mathbb{R}^d)$. By Theorem 3.6.16, $\mathcal{H}_2(\mathbb{R}^d)$ is the domain of the Laplacian Δ acting in $L^2(\mathbb{R}^d)$.

3.7 Exercises for Chapter 3

1. If $(\gamma_t, t \geq 0)$ is the heat kernel, give a direct proof (without using Itô's formula) that its characteristic function is

$$\phi_t(y) = e^{-t|y|^2},$$

   for all $y \in \mathbb{R}^d, t \geq 0$. Hint: Reduce the problem to the one-dimensional case, and then find and solve a differential equation involving ϕ_t.
2. Complete the proof that $\Phi_{c_t}(y) = e^{-t|y|}$ for all $y \in \mathbb{R}, t \geq 0$ by computing the relevant integral in the lower half-plane, and then dealing separately with the case $y = 0$. Hence show that $\int_{\mathbb{R}} c_t(x) dx = 1$.
3. Confirm that the binary operation of convolution of measures is both associative and commutative.
4. Show that if μ_1 and μ_2 are finite measures that are both absolutely continuous with respect to Lebesgue measure, with Radon–Nikodym derivatives f_1 and f_2 (respectively), then $\mu_1 * \mu_2$ is absolutely continuous with respect to Lebesgue measure with Radon–Nikodym derivative $f_1 * f_2$, which is the usual convolution of L^1-functions defined by

$$(f_1 * f_2)(x) = \int_{\mathbb{R}^d} f_1(x - y) f_2(y) dy,$$

   for Lebesgue almost all $x \in \mathbb{R}^d$. What can you say about absolute continuity of $\mu_1 * \mu_2$ with respect to Lebesgue measure when μ_2 is not absolutely continuous, but μ_1 remains so?
5. Derive the inequality (3.4.32). [Hint: Use the fact that for all $y \in \mathbb{R}$ there exists $0 < \theta < 1$ such that $e^{iy} - 1 - iy = \frac{\theta}{2} y^2$ (see Lemma 8.6 in Sato [84], p. 4.)].
6. It can be shown that for all $a \geq 0$, there exists a convolution semigroup $(\mu_t, t \geq 0)$ such that for all $t \geq 0, y \in \mathbb{R}^d$,

$$\Phi_{\mu_t}(y) = e^{-t(\sqrt{a^2 + |y|^2} - a)},$$

(see, e.g., Applebaum [6], pp. 41–2). Write down the generator of the associated C_0-semigroup on $S(\mathbb{R}^d)$ both as a pseudo-differential operator, and as a "function" of the Laplacian. This is in fact the well-known *relativistic Schrödinger operator*. What is special about the case $a = 0$?

7. Prove that $S(\mathbb{R}^d) \subseteq C_0(\mathbb{R}^d)$ and that $S(\mathbb{R}^d) \subseteq L^p(\mathbb{R}^d, \mathbb{R})$ for all $1 \leq p < \infty$. Hint: For $f \in S(\mathbb{R}^d)$, consider the function $x \to (1 + |x|^m)|f(x)|$, where $m \in \mathbb{N}$.

8. For each $\lambda > 0$, define the *resolvent kernel* U_λ of a convolution semigroup $(\mu_t, t \geq 0)$ by the prescription

$$U_\lambda(A) = \int_0^\infty e^{-\lambda t} \mu_t(A)dt,$$

for each $A \in \mathcal{B}(\mathbb{R}^d)$. (You may take the measurability of the mapping $t \to \mu_t(A)$ for granted.)[10]

(a) Show that U_λ is a finite measure.

(b) Show that for all $f \in C_0(\mathbb{R}^d)$, $x \in \mathbb{R}^d$,

$$R_\lambda f(x) = \int_0^\infty f(x + y)U_\lambda(dy).$$

(c) Show that if for all $t > 0$, μ_t has a density ρ_t, then U_λ has a density u_λ (called the λ-*resolvent density*) and given by

$$u_\lambda(x) = \int_0^\infty e^{-\lambda t} \rho_t(x)dt,$$

for Lebesgue almost all $x \in \mathbb{R}^d$.

(d) If $\int_0^\infty T_t f(x)dt$ exists for all $f \in C_0(\mathbb{R}^d)$ and $x \in \mathbb{R}^d$, then the semigroup is said to be *transient*.[11] In this case, under the conditions of (c), the *zero resolvent density* u_0 exists. It is known that the heat semigroup of Brownian motion of variance σ^2 is transient if and only if $d \geq 3$. Show that in this case, for $x \neq 0$ and $\sigma^2 = 1/2$, we obtain[12] the *Newtonian potential*

$$u_0(x) = \frac{\Gamma(d/2 - 1)}{2\pi^{d/2}} \frac{1}{|x|^{d-2}}.$$

---

[10] You can in fact prove this by approximating the indicator function of an open set by continuous functions and using weak continuity, then extend it to Borel sets by a monotone class argument.

[11] There are more probabilistic ways of describing this – see, e.g., Chapter 7 of Sato [84].

[12] I.e., replace the 4 by a 2 in the formula for the Gauss–Weierstrass function.

4

Self-Adjoint Semigroups and Unitary Groups

In this chapter we will mainly work with the case where E is a complex Hilbert space, which we denote by H. To simplify discussions, we will restrict to contraction semigroups within this context.

4.1 Adjoint Semigroups and Self-Adjointness

Let A be a linear operator in H with domain D_A. We will assume that it is densely defined, so that D_A is dense in H. If $g \in H$, we say that $g \in D_{A^*} :=$ Dom(A^*) if the mapping $f \to \langle Af, g \rangle$ from D_A to $\mathbb{C}$ is a bounded linear functional. Then, by density, it extends to a bounded linear functional on the whole of H, and so by the Riesz lemma there exists a unique $h \in H$ so that

$$\langle Af, g \rangle = \langle f, h \rangle.$$

We define a linear operator A^* in H with domain D_{A^*} by the prescription $A^*g = h$, and we call A^* the *adjoint* of A. It can be shown that A^* is in fact closed. Moreover D_{A^*} is dense in H if and only if A is closeable, in which case $\overline{A} = A^{**} := (A^*)^*$ (see, e.g., Reed and Simon [76], Theorem VIII.1, pp. 252–3I). It is a standard result in elementary Hilbert space theory that if A is bounded, then so is A^* with $||A^*|| = ||A||$.

Next, we consider adjoints of contraction semigroups in H.

Theorem 4.1.1 *If $(T_t, t \geq 0)$ is a contraction semigroup in H with generator A, then $(T_t^*, t \geq 0)$ is also a contraction semigroup having generator A^*.*

Proof. It is clear that T_t^* is a contraction for each $t \geq 0$, that $T_0^* = I$ and for all $s, t \geq 0$, $T_{s+t}^* = (T_t T_s)^* = T_s^* T_t^*$. For strong continuity, it is sufficient to prove $\lim_{t \to 0} T_t^* f = f$ for all $f \in H$. But

$$||T_t^* f - f||^2 = ||T_t^* f||^2 - 2\Re\langle T_t f, f\rangle + ||f||^2$$
$$\leq 2||f||^2 - 2\Re\langle T_t f, f\rangle \to 0 \text{ as } t \to 0,$$

and we have proved that $(T_t^*, t \geq 0)$ is a contraction semigroup. Let B denote its generator. We must prove that $B = A^*$, and then we are done. For $f \in D_A, g \in D_B$,

$$\langle Af, g\rangle = \lim_{t \to 0}\left\langle \frac{1}{t}(T_t - I)f, g\right\rangle$$
$$= \lim_{t \to 0}\left\langle f, \frac{1}{t}(T_t^* - I)g\right\rangle$$
$$= \langle f, Bg\rangle,$$

and so $B \subseteq A^*$.

Using (1.3.9), if $g \in D_{A^*}$, then for all $f \in D_A, t \geq 0$,

$$\langle f, T_t^* g - g\rangle = \langle T_t f - f, g\rangle$$
$$= \int_0^t \langle AT_s f, g\rangle ds$$
$$= \int_0^t \langle f, T_s^* A^* g\rangle ds,$$

so by density of D_A in H, we conclude that

$$T_t^* g - g = \int_0^t T_s^* A^* g ds.$$

Dividing both sides of the last identity by t and then taking the limit as $t \to 0$, we find using (RI4) that $Bg = A^* g$, hence $A^* \subseteq B$, and the proof is complete. □

A densely defined linear operator X with domain D_X is said to be *symmetric* if $X \subseteq X^*$, i.e., $\langle X\psi_1, \psi_2\rangle = \langle \psi_1, X\psi_2\rangle$ for all $\psi_i \in D_X, i = 1, 2$. The operator X is said to be *self-adjoint* if $X = X^*$, i.e., it is symmetric and $D_{X^*} = D_X$. A symmetric operator X may have many or no self-adjoint extensions. We say that X is *essentially self-adjoint* if it has a unique self-adjoint extension, i.e., $\overline{X} = X^*$. Of these three related types of operators, it is only the self-adjoint ones to which the *spectral theorem* applies, i.e., there exists a projection valued measure $E: \mathcal{B}(\mathbb{R}) \to \mathcal{P}(H)$, where $\mathcal{P}(H)$ is the lattice of all orthogonal projections acting in H, such that

$$X = \int_{\mathbb{R}} \lambda \, dE\lambda,$$

so that for all $\phi \in D_X$, $\psi \in H$,

$$\langle X\phi, \psi \rangle = \int_{\mathbb{R}} \lambda d \langle \phi, E(\lambda)\psi \rangle,$$

where the right-hand side of the last display is a standard Lebesgue integral with respect to the complex measure[1] $\langle \phi, E(\cdot)\psi \rangle$. We also have the characterisation of the domain

$$D_X = \left\{ \phi \in H; \int_{\mathbb{R}} \lambda^2 d||E(\lambda)\phi||^2 < \infty \right\}.$$

We have an associated *functional calculus* in that if $g: \mathbb{R} \to \mathbb{R}$ is a Borel function, then $g(X)$ is self-adjoint where

$$g(X) = \int_{\mathbb{R}} g(\lambda)dE\lambda,$$

with domain,

$$D_{g(X)} = \left\{ \phi \in H; \int_{\mathbb{R}} |g(\lambda)|^2 d||E(\lambda)\phi||^2 < \infty \right\}.$$

If g is bounded, then $D_{g(X)} = H$ and $g(X)$ is a bounded self-adjoint operator. If h is also bounded, then $(gh)(X) = g(X)h(X) = h(X)g(X)$ is again bounded and self-adjoint. If X is self-adjoint, then its spectrum $\sigma(X) \subseteq \mathbb{R}$. For details, see, e.g., section VIII.3 in Reed and Simon [76], section VIII.3, pp. 259–64, or Akhiezer and Glazman [3], section 75, pp. 247–53.

We return to the study of semigroups, and first note that if a contraction semigroup $(T_t, t \geq 0)$ is self-adjoint for all $t \geq 0$, then it is positive for all $t \geq 0$, as for all $f \in H$ we have

$$\langle T_t f, f \rangle = ||T_{t/2} f||^2 \geq 0.$$

Corollary 4.1.2 *The contraction semigroup $(T_t, t \geq 0)$ is self-adjoint if and only if its generator A is self-adjoint.*

Proof. If $T_t = T_t^*$ for all $t \geq 0$, then $A = A^*$ by Theorem 4.1.1. Conversely, if $A = A^*$, then for all $t \geq 0$, using functional calculus,

$$T_t^* = e^{tA^*} = e^{tA} = T_t. \qquad \square$$

4.1.1 Positive Self-Adjoint Operators

We have seen this idea before, but it will be helpful to make it more precise. Let B be a positive, self-adjoint operator on a Hilbert space H. Define

---

[1] It is a finite signed measure if H is a real Hilbert space.

$A = -B$. Then it is immediate that A is dissipative. Furthermore it is known that the spectrum $\sigma(A) \subseteq (-\infty, 0]$, and so for all $\lambda > 0$, $\text{Ran}(\lambda I - A) = H$. Hence by Theorem 2.2.6 (the Lumer–Phillips theorem), A is the generator of a contraction semigroup $(T_t, t \geq 0)$. We must show that it is self-adjoint. In fact we have

Theorem 4.1.3 *If B is a positive, self-adjoint operator on a Hilbert space, then $A = -B$ is the generator of a self-adjoint contraction semigroup $(T_t, t \geq 0)$. Furthermore if $B = \int_0^\infty \lambda \, dE(\lambda)$, then for all $t \geq 0$,*

$$T_t = e^{-tB} = \int_0^\infty e^{-t\lambda} dE(\lambda).$$

Proof. For each $t \geq 0$, define $S_t = e^{-tB} = \int_0^\infty e^{-t\lambda} dE(\lambda)$. Then S_t is self-adjoint, and a contraction as for all $\psi \in H$,

$$||S_t \psi||^2 = \int_0^\infty e^{-2t\lambda} ||dE(\lambda)\psi||^2 \leq ||\psi||^2.$$

We also have that $(S_t, t \geq 0)$ is a semigroup; (S1) and (S2) follow from functional calculus, and for strong continuity we have

$$\lim_{t \to 0} ||S_t \psi - \psi||^2 = \lim_{t \to 0} \int_0^\infty (e^{-2t\lambda} - 1)^2 ||dE(\lambda)\psi||^2 = 0,$$

by dominated convergence. Let X be the infinitesimal generator of $(S_t, t \geq 0)$. Then $A \subseteq X$ as for all $\psi \in D_B$, and for some $0 < \theta < 1$,

$$\lim_{t \to 0} \left|\left| \frac{S_t \psi - \psi}{t} + B\psi \right|\right|^2$$

$$= \lim_{t \to 0} \int_0^\infty \left(\frac{e^{-t\lambda} - 1}{t} + \lambda \right)^2 ||dE(\lambda)\psi||^2$$

$$= \lim_{t \to 0} t \int_0^\infty \lambda^2 e^{-2t\theta\lambda} ||dE(\lambda)\psi||^2$$

$$\leq \lim_{t \to 0} t ||B\psi||^2 = 0.$$

The fact that $A = X$ now follows by the argument at the end of the proof of Theorem 2.2.4. So by uniqueness of generators, we have $S_t = T_t$ for all $t \geq 0$, and the result follows. $\square$

We have a converse to the last result.

Theorem 4.1.4 *If the linear operator $-B$ is the generator of a self-adjoint contraction semigroup in a Hilbert space H, then B is positive and self-adjoint.*

Proof. (Based on Davies [25], Theorem 4.6, pp. 99–100). Assume that $(T_t, t \geq 0)$ is a self-adjoint contraction semigroup. Then, using Theorem 1.5.25(2), we see that the resolvent $R_\lambda = \int_0^\infty e^{-\lambda t} T_t dt$ is a self-adjoint contraction for all $\lambda > 0$. But then $(I + B)^{-1}$ is a self-adjoint contraction. By functional calculus (or Corollary 4.1.2),

$$B = \frac{I}{(I + B)^{-1}} - I$$

is self-adjoint. But the spectrum of $\frac{I}{(I+B)^{-1}}$ is contained in $(1, \infty)$, hence $\sigma(B) \subseteq (0, \infty)$, and so B is positive, as required. $\qquad\square$

Now suppose that B is a densely defined, symmetric positive operator. Then it has an extension B_F, called the *Friedrichs extension*, which is positive self-adjoint. So we may conclude from the above that $-B_F$ is the generator of a self-adjoint contraction semigroup on H.

So we have the very useful fact that the negation of every symmetric positive operator has a self-adjoint extension that generates a self-adjoint contraction semigroup.

As the Friedrichs extension is so important, we present a proof, based on the highly accessible account in Baudoin [10], pp. 99–100 (see also, e.g., Reed and Simon [77], Theorem X.23, p. 177).

Theorem 4.1.5 (Friedrichs) *If B is a densely defined symmetric positive linear operator in a Hilbert space H, then it has a self-adjoint positive extension B_F.*

Proof. Define the following bilinear form on D_B:

$$\langle \phi, \psi \rangle_1 = \langle \phi, \psi \rangle + \langle B\phi, \psi \rangle,$$

for all $\phi, \psi \in D_B$. It is easily verified that $\langle \cdot, \cdot \rangle_1$ is an inner product on D_B. We complete D_B in the associated norm to obtain a Hilbert space H_1, in which D_B is, of course, dense. Since $||\phi|| \leq ||\phi||_1$ for all $\phi, \psi \in D_B$, we see that the mapping $\iota(f) = f$ from $D_B \subseteq H_1$ to H is a linear contraction, and so extends to a contraction from H_1 to H. We show that ι is injective. To see this, assume that $\phi \in H_1$ satisfies $\iota(\phi) = 0$. Then we can find a sequence (ϕ_n) in D_A which converges to ϕ in H_1 as $n \to \infty$. Since $\iota(\phi) = 0$, we must have $\lim_{n\to\infty} ||\phi_n|| = 0$. Then it follows that

$$||\phi||_1 = \lim_{m,n\to\infty} \langle \phi_m, \phi_n \rangle + \lim_{m,n\to\infty} \langle B\phi_m, \phi_n \rangle = 0,$$

so ι is injective. Define $S = \iota \circ \iota^*$. Then it is a self-adjoint bounded linear operator in H. We next show that S is injective. Since ι is injective, if $\phi \in H$

with $S\phi = 0$ then $\iota^*\phi = 0$. Hence for all $\psi \in H_1$,

$$0 = \langle \iota^*\phi, \psi \rangle_1 = \langle \phi, \iota(\psi) \rangle.$$

But then ψ is orthogonal to $\mathrm{Ran}(\iota) \supseteq D_B$, which is dense in H. So $\psi = 0$, as required. Now it is a fact that every injective bounded self-adjoint operator in a Hilbert space has a densely defined self-adjoint inverse. See Lemma A.6 on p. 259 of Baudoin [10] for an elegant proof of this. Then, for all $\phi, \psi \in D_B$, we have

$$
\begin{aligned}
\langle \phi, \psi \rangle + \langle B\phi, \psi \rangle &= \langle \iota^{-1}(\phi), \iota^{-1}(\psi) \rangle_1 \\
&= \langle ((\iota^{-1})^* \circ \iota^{-1})\phi, \psi \rangle \\
&= \langle (\iota \circ \iota^*)^{-1}\phi, \psi \rangle \\
&= \langle S^{-1}\phi, \psi \rangle.
\end{aligned}
$$

So $D_B \subseteq D_{S^{-1}}$. Furthermore $S^{-1} - I$ is a self-adjoint extension of B, and is positive as

$$\langle (S^{-1} - I)\phi, \phi \rangle = \langle B\phi, \phi \rangle \geq 0.$$

So we may take $B_F := S^{-1} - I$, and the result is established. $\qquad\square$

4.1.2 Adjoints of Semigroups on Banach Spaces

We briefly remark about the extension of some of the main results of this section to a general Banach space E. Let X be a densely defined linear operator in E with domain D_X. If $\phi \in E'$, we say that $\phi \in D_{X^*} := \mathrm{Dom}(X^*)$ if there exists $\psi \in E'$ so that for all $f \in D_X$

$$\langle Xf, \phi \rangle = \langle f, \psi \rangle.$$

We then define the adjoint X^* of X to be the (unique) operator in E' with domain D_{X^*} for which $X^*\phi = \psi$. It is easily seen that X^* is linear, and is bounded if X is.

Recall that a Banach space E is *reflexive* if E^{**} and E are isometrically isomorphic. All Hilbert spaces are reflexive. If (S, Σ, μ) is a measure space, then $L^p(S, \Sigma, \mu)$ is reflexive if and only if $1 < p < \infty$. We state without proof the following extension of Theorem 4.1.1 to reflexive Banach spaces.

Theorem 4.1.6 *If $(T_t, t \geq 0)$ is a C_0-semigroup on a reflexive Banach space E, then $(T_t^*, t \geq 0)$ is a C_0-semigroup on E'.*

A proof can be found in section 7.3 of Davies [27] pp. 197–8. The main difficulty, in comparison with the proof of Theorem 4.1.1, is establishing strong continuity (S3). Note that it is easy to prove weak continuity at $t = 0$, i.e., that $\lim_{t \to 0} \langle f, T_t^* \phi \rangle = \langle f, \phi \rangle$, for all $f \in E, \phi \in E'$, since

$$\lim_{t \to 0} \langle f, T_t^* \phi \rangle = \lim_{t \to 0} \langle T_t f, \phi \rangle = \langle f, \phi \rangle,$$

and one approach to the problem is to establish that weak continuity at zero implies strong continuity at zero. That route is taken by Sinha and Srivasta [91], p. 37.

4.2 Self-Adjointness and Convolution Semigroups

In this section we return to the study of convolution semigroups $(\mu_t, t \geq 0)$ on $\mathbb{R}^d$, and we consider the associated contraction semigroup $(T_t, t \geq 0)$ acting in $L^2(\mathbb{R}^d)$. To each μ_t, we may associate the *reversed measure* $\widetilde{\mu}_t$ defined by $\widetilde{\mu}_t(A) = \mu_t(-A)$ for all $A \in \mathcal{B}(\mathbb{R}^d)$. It is an easy exercise to check that $(\widetilde{\mu}_t, t \geq 0)$ is also a convolution semigroup in $\mathbb{R}^d$, and it has an associated contraction semigroup $(\widetilde{T}_t, t \geq 0)$ acting in $L^2(\mathbb{R}^d)$. Then for all $t \geq 0, f \in L^2(\mathbb{R}^d), x \in \mathbb{R}^d$,

$$\widetilde{T}_t f(x) = \int_{\mathbb{R}^d} f(x+y) \widetilde{\mu}_t(dy) = \int_{\mathbb{R}^d} f(x-y) \mu_t(dy). \tag{4.2.1}$$

Proposition 4.2.7 *For all $t \geq 0$, $\widetilde{T}_t = T_t^*$.*

Proof. For all $f, g \in L^2(\mathbb{R}^d)$, using Fubini's theorem

$$\begin{aligned} \langle \widetilde{T}_t f, g \rangle &= \int_{\mathbb{R}^d} \int_{\mathbb{R}^d} f(x-y) g(x) \mu_t(dy) dx \\ &= \int_{\mathbb{R}^d} \int_{\mathbb{R}^d} f(x) g(x+y) \mu_t(dy) dx \\ &= \langle f, T_t g \rangle, \end{aligned}$$

and the result follows. $\square$

We say that a convolution semigroup is *symmetric* if $\mu_t = \widetilde{\mu}_t$ for all $t \geq 0$. In this case we see immediately from Proposition 4.2.7 that the associated contraction semigroup is self-adjoint in $L^2(\mathbb{R}^d)$. Much more is true, as we see from the following theorem, which is the main result of this section. It is based on Theorem 5.4.1 in Applebaum [7], pp. 140–2.

Theorem 4.2.8 *The following are equivalent.*

(i) *The convolution semigroup $(\mu_t, t \geq 0)$ is symmetric.*
(ii) *$T_t = T_t^*$ for each $t \geq 0$.*

(iii) $A = A^*$.

(iv) $\Im(\eta) = 0$.

(v) $b = 0$, $v = \tilde{v}$.

(vi) For all $f \in S(\mathbb{R}^d)$,

$$\mathcal{A}f(x) = \sum_{i,j=1}^{d} a_{ij}\partial_i\partial_j f(x) + \frac{1}{2}\int_{\mathbb{R}^d} (f(x+y) - 2f(x) + f(x-y))v(dy).$$

$$(4.2.2)$$

Proof. (i) $\Rightarrow$ (ii) has just been shown. To see that (ii) $\Rightarrow$ (i) observe that for all $A \in \mathcal{B}(\mathbb{R}^d)$ by (3.3.18),

$$\mu_t(A) = T_t 1_A(0) = T_t^* 1_A(0) = \tilde{T}_t 1_A(0) = \tilde{\mu}_t(A).$$

(ii) $\Leftrightarrow$ (iii) is Corollary 4.1.2.

To verify that (i) $\Rightarrow$ (iv), we have for all $y \in \mathbb{R}^d$, $\Phi_{\mu_t}(y) = e^{t\eta(y)}$ and $\Phi_{\tilde{\mu}_t}(y) = e^{t\overline{\eta(y)}}$. Now take logarithms to obtain $\eta(y) = \overline{\eta(y)}$, as required.

(iv) $\Rightarrow$ (i) follows from the fact that the characteristic function uniquely determines the measure.

(iv) $\Rightarrow$ (v). As we have shown that (i) to (iv) are equivalent, we may assume (i), and then to show that $v = \tilde{v}$ we use a formula that is derived in Sato [84], Corollary 8.9, p. 45:

$$\int_{\mathbb{R}^d} f(y)v(dy) = \lim_{t \to 0} \frac{1}{t}\int_{\mathbb{R}^d} f(y)\mu_t(dy),$$

for all bounded continuous functions that are bounded away from zero. It follows from this that $\int_{\mathbb{R}^d}(\sin(x \cdot y) - 1_{B_1}(y)x \cdot y)v(dy) = 0$, and hence $b = 0$ as required.

(v) $\Rightarrow$ (iv) is obvious.

(iii) $\Rightarrow$ (vi) For simplicity, assume $a = 0$ (the reader should easily be able to deal with the general case). Then, for each $f \in S(\mathbb{R}^d)$,

$$Af = \int_{\mathbb{R}^d} (f(x+y) - f(x) - \sum_{i=1}^{d} y^i \partial_i f(x)1_{|y|<1}(y))v(d\tau)$$

and

$$A^*f = \int_{\mathbb{R}^d} (f(x-y) - f(x) + \sum_{i=1}^{d} y^i \partial_i f(x)1_{|y|<1}(y))v(d\tau).$$

The result follows by using the identity $Af = \frac{1}{2}(Af + A^*f)$.

Finally, (vi) $\Rightarrow$ (iv) follows from rewriting the operator $\mathcal{A}$ as a pseudo-differential operator, as in (3.6.41). Then, for each $f \in S(\mathbb{R}^d), x \in \mathbb{R}^d$, we have

$$Af(x) = (2\pi)^{-d/2}\frac{1}{2}\int_{\mathbb{R}^d}\left[e^{iu\cdot(x+y)} - 2e^{iu\cdot x} + e^{iu\cdot(x-y)}\right]\hat{f}(u)du$$

$$= (2\pi)^{-d/2}\int_{\mathbb{R}^d}e^{iu\cdot x}(\cos(u\cdot y) - 1)\hat{f}(u)du.$$

From this and (3.3.19), it follows that $\eta(u) = \cos(u\cdot y) - 1 \in \mathbb{R}$ for all $u, y \in \mathbb{R}^d$. Here we have again taken $a = 0$ for convenience. $\square$

The proof of Theorem 4.2.8 has also shown that the symbol of the self-adjoint generator A has the generic form

$$\eta(y) = -ay\cdot y - \int_{\mathbb{R}^d}(1 - \cos(u\cdot y))v(dy) \leq 0,$$

for all $y \in \mathbb{R}^d$. From this, we may use Parseval's identity to show that $-A$ is a positive operator, since by Theorem 3.6.16, for all $f \in D_A = \mathcal{H}_\eta(\mathbb{R}^d)$,

$$-\langle Af, f\rangle = -\langle\widehat{Af}, \hat{f}\rangle = -\int_{\mathbb{R}^d}\eta(y)|\hat{f}(y)|^2dy \geq 0.$$

4.3 Unitary Groups, Stone's Theorem

Semigroups describe irreversible phenomena. For reversible phenomena we need groups. The definition of a C_0-group in a Banach space E parallels that of a semigroup:

A C_0-*group* on E is a family of bounded, linear operators $(V_t, t \in \mathbb{R})$ on E for which

1. $V_{s+t} = V_s V_t$ for all $s, t \in \mathbb{R}$.
2. $V_0 = I$.
3. The mapping $t \to V_t\psi$ from $\mathbb{R}$ to E is continuous for all $\psi \in E$.

Observe that since for all $t \in \mathbb{R}, V_{-t}V_t = V_tV_{-t} = I$, each operator V_t is invertible, with $V_t^{-1} = V_{-t}$.

The theory of C_0-groups is fairly similar to that of semigroups and we will not pursue it in detail. A lot of results can be found in the literature, and we have already encountered some of these in Problems 2.5 and 2.6. The notion of an infinitesimal generator is defined in exactly the same way as for semigroups, the only difference being that the limit as $t \to 0$ is taken in the topology of $\mathbb{R}$, and not of $[0, \infty)$. We prove only one general result. For that we need the following lemma.

Lemma 4.3.9 *If* $(T_t, t \geq 0)$ *is a* C_0-*semigroup in* E *and* $t \to g(t)$ *is a mapping from* $[0, \infty)$ *to* E *so that* $\lim_{t \to 0} g(t)$ *exists and equals* ψ *(say), then*

$$\lim_{t \to 0} T_t g(t) = \psi.$$

Proof. Using (1.2.4) we have

$$\begin{aligned}
\|T_t g(t) - \psi\| &\leq \|T_t\| . \|g(t) - \psi\| + \|T_t \psi - \psi\| \\
&\leq M e^{at} \|g(t) - \psi\| + \|T_t \psi - \psi\| \\
&\to 0 \text{ as } t \to 0.
\end{aligned}$$ $\square$

The next result is taken from Pazy [74] pp. 23–4.

Proposition 4.3.10 *If* $(T_t, t \geq 0)$ *is a* C_0-*semigroup in* E *with generator* A *which is such that* T_t *has a bounded inverse for all* $t > 0$, *then* $(T_t^{-1}, t \geq 0)$ *is a* C_0-*semigroup having generator* $-A$. *Furthermore* $(V_t, t \in \mathbb{R})$ *is a* C_0-*group with generator* A, *where*

$$V_t = \begin{cases} T_t & \text{if } t \geq 0, \\ T_{-t}^{-1} & \text{if } t < 0. \end{cases} \tag{4.3.3}$$

Proof. To show that $(T_t^{-1}, t \geq 0)$ is a C_0-semigroup, note first that for $s, t \geq 0$, we have

$$T_{s+t}^{-1} = (T_t T_s)^{-1} = T_s^{-1} T_t^{-1}.$$

For strong continuity, since T_t is surjective for all $t > 0$, given any $g \in E$ there exists $f \in E$ so that $g = T_t f$. Then, for $s < t$, we have

$$\|T_s^{-1} g - g\| = \|T_{t-s} f - T_t f\| \to 0 \text{ as } s \to 0.$$

Now let B be the generator of $(T_t^{-1}, t \geq 0)$. Then for $f \in D_B$, by Lemma 4.3.9, we have

$$Bf = \lim_{t \to 0} \frac{T_t^{-1} f - f}{t} = \lim_{t \to 0} T_t \left(\frac{T_t^{-1} f - f}{t} \right) = \lim_{t \to 0} \frac{f - T_t f}{t}.$$

Hence $f \in D_A$ and $B \subseteq -A$. A similar argument shows that $-A \subseteq B$, and hence $B = -A$, as required.

To see that $(V_t, t \in \mathbb{R})$ is a C_0-group with generator A is fairly straightforward, and we only include a few pointers here: if $t > s > 0$ then

$$V_t V_{-s} = T_t T_s^{-1} = T_{t-s} = V_{t-s},$$

and if $f \in D_A$,

$$\lim_{t \uparrow 0} \frac{V_t f - f}{t} = \lim_{t \uparrow 0} \frac{T_{-t}^{-1} f - f}{t} = -\lim_{s \downarrow 0} \frac{T_s^{-1} f - f}{s} = Af.$$ $\square$

The notion of convolution group of measures is defined by an obvious extension of that of semigroups. To the author's knowledge, there is only one class of examples, indexed by $b \in \mathbb{R}^d$, and taking the form $\{\delta_{bt}, t \in \mathbb{R}\}$. This induces the translation group on $C_0(\mathbb{R}^d)$ and $L^p(\mathbb{R}^d)$ for $1 \leq p < \infty$.

A very important example of a C_0-group is obtained by taking E to be a Hilbert space and V_t, which, as is conventional, we write as U_t, to be a unitary operator[2] for all $t \in \mathbb{R}$. We then call $(U_t, t \in \mathbb{R})$ a *one-parameter unitary group*. Note that for all $t \in \mathbb{R}$,

$$U_{-t} = U_t^{-1} = U_t^*.$$

If in Proposition 4.3.10, we take each T_t to be a unitary operator in H, then the bounded inverse hypothesis is automatically satisfied and $(V_t, t \geq 0)$, as given by (4.3.3), is a one-parameter unitary group. Conversely, any one-parameter unitary group is of the form (4.3.3), with each operator therein being unitary.

One of the reasons why one-parameter unitary groups are so important is the role that they play in *quantum mechanics*, so let us take a quick excursion to make the necessary connection.[3] The basic equation of (nonrelativistic) quantum mechanics is the *Schrödinger equation* which describes the infinitesimal change in a curve $(\psi(t), t \in \mathbb{R})$ in the complex Hilbert space $L^2(\mathbb{R}^d; \mathbb{C})$ (where we would take $d = 3$ if we want to do real-world physics):

$$i\hbar \frac{\partial \psi}{\partial t} = \mathcal{H}\psi. \qquad (4.3.4)$$

Here $i = \sqrt{-1}$ and $\hbar = h/2\pi$, where h is Planck's constant, whose value is approximately 6.626×10^{-34} in units of joule-seconds. $\mathcal{H}$ is a self-adjoint operator[4] called the *Hamiltonian*. It represents the total energy in the quantum system and it takes the form:

$$\mathcal{H} = -\frac{\hbar^2}{2m}\Delta + V, \qquad (4.3.5)$$

where V is a multiplication operator called the *potential*. For example, in an electromagnetic interaction, V would be multiplication by $1/|x|$ ($x \neq 0$). The right-hand side of (4.3.5) represents the sum of kinetic and

---

[2] A bounded linear operator in H is *unitary* if it an isometry and a co-isometry, i.e.,

$$UU^* = U^*U = I.$$

Equivalently, U is unitary if it is invertible with $U^{-1} = U^*$.

[3] For a solid mathematical introduction to this subject, see Hannabus [43].

[4] The physical principles of quantum mechanics impose self-adjointness on the unbounded operator H. To be symmetric is not enough. For more background on this, see, e.g., Mackey [64].

potential energies. To learn more about the Schrödinger equation, we will study an abstract version of it in an arbitrary complex Hilbert space H. So we investigate

$$\frac{\partial \psi}{\partial t} = iX\psi(t), \tag{4.3.6}$$

where X is self-adjoint, and observe that the operator $A = iX$, which is the natural candidate to be the generator of a C_0-group implementing the solution to (4.3.6), is skew-adjoint, i.e., $A = -A^*$.

The connection between one parameter unitary groups and abstract Schrödinger equations is given by *Stone's theorem* which we will now prove, using the approach of Goldstein [40] Theorem 4.7, pp. 32–3. This gives us an opportunity to employ results that were established in Chapter 2.

Theorem 4.3.11 (Stone's Theorem) *A densely defined linear operator A in H is the generator of a one-parameter unitary group if and only if it is skew-adjoint.*

Proof. If A is the generator of the one-parameter unitary group $(U_t, t \in \mathbb{R})$, then for all $f \in H$,

$$\frac{1}{t}(U_{-t}f - f) = \frac{1}{t}(U_t^* f - f).$$

It follows that $D_A = D_{A^*}$ and, taking limits as $t \to 0$ in the last display with $f \in D_A$, we get $A^* = -A$.

Conversely, suppose that A is skew-adjoint. Then it is closed and for all $f \in D_A$,

$$\langle Af, f \rangle = \langle f, A^* f \rangle = -\langle f, Af \rangle = -\overline{\langle Af, f \rangle}.$$

Hence $\Re\langle Af, f \rangle = 0$, and so $\pm A$ is dissipative. By the Lumer–Phillips theorem (Theorem 2.2.7), we know that $\pm A$ are generators of contraction semigroups if there exists $\lambda > 0$ so that $\mathrm{Ran}(\lambda I \pm A)$ is dense in H. Let $g \in H$ with $g \perp \mathrm{Ran}(\lambda I \pm A)$ for all $\lambda > 0$. Then, for all $f \in D_A$,

$$\langle (\lambda I \pm A)f, g \rangle = 0.$$

Then, $g \in D_{A^*} = D_A$ and by density of D_A in H, we have $(\lambda I \mp A^*)g = 0$, i.e.,

$$-Ag = A^* g = \pm \lambda g.$$

It follows that $\Re\langle Ag, g \rangle = \mp \lambda \|g\|^2$. But we know $\Re\langle Ag, g \rangle = 0$ and so $g = 0$. Hence $(e^{tA}, t \geq 0)$ and $(e^{-tA}, t \geq 0)$ are contraction semigroups in H, and the hypothesis of Proposition 4.3.10 is clearly satisfied. As in (4.3.3), we define $(U_t, t \in \mathbb{R})$ by

$$U_t = \begin{cases} e^{tA} & \text{if } t \geq 0, \\ e^{(-t)(-A)} & \text{if } t < 0. \end{cases}$$

Then, $(U_t, t \in \mathbb{R})$ is a C_0 contraction group, so for all $t \in \mathbb{R}, \psi \in H, ||U_t\psi|| \leq ||\psi||$. But then

$$||U_{-t}U_t\psi|| = ||\psi|| \leq ||U_t\psi||,$$

and so U_t is an isometry. But we know it is invertible, with $U_t^{-1} = U_{-t}$. It follows that U_t is unitary, and we are done. $\qquad\qquad\qquad\square$

It follows from Stone's theorem (and a suitable extension of Theorem 1.3.15 to groups) that the unique solution to the abstract Schrödinger equation (4.3.6), with initial condition $\psi(0) = \psi \in H$, is given by $\psi(t) = U_t\psi$, for all $t \geq 0$, where $U_t = e^{itX}$ gives rise to a one-parameter unitary group. Note also that the unitarity ensures that $||\psi(t)|| = ||\psi||$ for all $t \in \mathbb{R}$, so the dynamics of the Schrödinger equation change the amplitude, but not the modulus, of the vector ψ. If X has spectral decomposition $\int_{-\infty}^{\infty} \lambda dE(\lambda)$, then, by functional calculus, for each $t \in \mathbb{R}$, $U_t = \int_{-\infty}^{\infty} e^{it\lambda} dE(\lambda)$.

Returning to (4.3.5), it is clearly important to know when the operator $X = \Delta + V$ is, at least, essentially self-adjoint so that Stone's theorem, applied to its unique self-adjoint extension, gives rise to a legitimate dynamics. For some insight into this problem, see, e.g., section X.2 of Reed and Simon [77], pp. 162–76, where you can find information about essential self-adjointness of (non-relativistic) atomic Hamiltonians, which incorporate Coulombic potentials that describe the electron–nucleus attractions, and electron–electron repulsions within the atom. For this you need results of the following kind: if $V \in L^2(\mathbb{R}^3, \mathbb{R}) + L^\infty(\mathbb{R}^3, \mathbb{R})$, then[5] $-\Delta + V$ is essentially self-adjoint on the domain $C_c^\infty(\mathbb{R}^3; \mathbb{C})$, and self-adjoint on D_Δ. This result includes the *free Hamiltonian*, when $V = 0$.

In quantum theory, the unit vectors $\psi \in H$ describe the *state* or environment of a quantum system, and the dynamics that we have just described, which give the evolution of states, is called the *Schrödinger picture* by physicists. On the other hand self-adjoint operators in H (typically unbounded ones) describe *observables* in quantum theory, such as position, momentum or total energy (which as we have seen, is represented by the Hamiltonian operator). A one-parameter unitary group $(U_t, t \in \mathbb{R})$ also gives rise to a dynamical evolution of observables, and this is called the *Heisenberg picture* by physicists. For

---

[5] The notation indicates that V takes the form $V_1 + V_2$, where $V_1 \in L^2$ and $V_2 \in L^\infty$.

simplicity, assume that $X = X^*$ is a bounded self-adjoint operator in H; then, since for all $t \in \mathbb{R}$, $\psi \in H$,

$$U_t X \psi = U_t X U_t^{-1} U_t \psi = U_t X U_t^* \psi(t),$$

we see that the dynamical evolution of the observable X is given by $X \to U_t X U_t^*$, for $t \in \mathbb{R}$. We thus define $(j_t, t \geq 0)$ to be the one-parameter group of $*$–automorphisms of $\mathcal{L}(H)$ given by

$$j_t(X) = U_t X U_t^*,$$

for each $t \in \mathbb{R}$, $X \in \mathcal{L}(H)$. So (as is easily verified) we have $j_{s+t} = j_s j_t$, for all $s, t \in \mathbb{R}$, and $j_t(X)^* = j_t(X^*)$, for each $t \in \mathbb{R}$, $X \in \mathcal{L}(H)$.

We would like to derive a differential equation that expresses the infinitesimal change in j_t. For simplicity, we will assume that $A \in \mathcal{L}(H)$ and argue fairly informally. Then, the Leibniz rule applied to (4.3.6) gives

$$\frac{dj_t(X)}{dt} = \frac{dU_t}{dt} X U_t^* + U_t X \frac{dU_t^*}{dt}.$$

Taking formal adjoints in (4.3.6), with $\psi(t) = U_t \psi$, yields

$$\frac{dU_t^*}{dt} = -i U_t^* A,$$

and so we obtain the *Heisenberg equation of motion*

$$\frac{dj_t(X)}{dt} = j_t(i[A, X]), \tag{4.3.7}$$

where $[A, X] := AX - XA$ is the usual *commutator bracket*. We define $\delta : \mathcal{L}(H) \to \mathcal{L}(H)$ by $\delta(X) = i[A, X]$. Then, for all $X, Y \in \mathcal{L}(H)$,

$$\begin{aligned}
\delta(XY) &= i[A, XY] \\
&= i(AXY - XYA) \\
&= i(AXY - XAY + XAY - XYA) \\
&= i[A, X]Y + X.i[A, Y] \\
&= \delta(X)Y + X\delta(Y).
\end{aligned}$$

In general, linear mappings between algebra that satisfy the abstract Leibniz rule that we have just derived are called *derivations*. A derivation that is obtained by commutator bracket with a fixed element of the algebra (as above) is called an *inner derivation*. The study of derivations is an important theme in the exploration of dynamical systems on C^*-algebras, and applications to quantum statistical mechanics (see, e.g., Bratteli [18]).

4.4 Quantum Dynamical Semigroups

All of the material in the last section deals with reversible dynamics in quantum theory. To study irreversible processes at the quantum level, we need to introduce *quantum dynamical semigroups*. The natural context for this is an operator algebraic framework, but for simplicity let us continue to work on the algebra $\mathcal{L}(\mathcal{H})$ of all bounded linear operators acting on the complex Hilbert space $\mathcal{H}$. Then, a quantum dynamical semigroup is a contraction semigroup $(T_t, t \geq 0)$ defined on $\mathcal{L}(\mathcal{H})$ that has two additional properties:

1. (Conservativity) $T_t(I) = I$ for all $t \geq 0$.
2. (Complete Positivity) For all $X_1, \ldots, X_n, Y_1, \ldots, Y_n \in \mathcal{L}(\mathcal{H}), n \in \mathbb{N}$,
   $\sum_{i=1}^{n} Y_i^* T_t(X_i^* X_j) Y_j$ is a positive self-adjoint operator.

Of these two conditions, (1) is quite standard and we will return to it later. (2) is more surprising; it comes from physical considerations. We have the following classification result.

Theorem 4.4.12 (Gorini, Kossakowski, Sudarshan, Lindblad) *Let $\mathcal{H}$ be a complex, separable Hilbert space. A linear operator A on $\mathcal{L}(\mathcal{H})$ is the generator of a norm-continuous quantum dynamical semigroup if and only if there exists $H = H^* \in \mathcal{L}(\mathcal{H})$ and a sequence $(L_n, n \in \mathbb{N})$ of operators in $\mathcal{L}(\mathcal{H})$ for which $\sum_{n=1}^{\infty} L_n^* L_n$ is strongly convergent so that for all $X \in \mathcal{L}(\mathcal{H})$,*

$$A(X) = i[H, X] - \frac{1}{2} \sum_{n=1}^{\infty} (L_n^* L_n X - 2L_n^* X L_n + X L_n^* L_n), \tag{4.4.8}$$

where the right-hand side of the above equation is a strongly convergent series. We direct the reader to Parthasarathy [73] pp. 267–8 for the proof. First, let us be clear that to say that $\sum_{n=1}^{\infty} L_n^* L_n$ is strongly convergent, means that for any $\psi \in \mathcal{H}$, the sequence whose Nth term is $\sum_{n=1}^{N} L_n^* L_n \psi$ converges to $T\psi$ in $\mathcal{H}$ as $N \to \infty$, where $T \in \mathcal{L}(\mathcal{H})$, and by convention we write $T = \sum_{n=1}^{\infty} L_n^* L_n$. Secondly, for (4.4.8) to have meaning, we also need strong convergence of the sequence whose Nth term is $\sum_{n=1}^{N} L_n^* X L_n \psi$ for any $X \in \mathcal{L}(\mathcal{H})$. The fact that we do not need to assume this as well follows from the next lemma, the proof of which was communicated to me by Alex Belton.

Lemma 4.4.13 *If $\sum_{n=1}^{\infty} L_n^* L_n$ is strongly convergent, then for all $X \in \mathcal{L}(\mathcal{H})$, $\sum_{n \in \mathbb{N}} L_n^* X L_n$ is strongly convergent to a bounded linear operator in $\mathcal{L}(\mathcal{H})$.*

Proof. Let K be another complex separable Hilbert space in which we choose a complete orthonormal basis (e_n). We first show that $\sum_{n \in \mathbb{N}} L_n x \otimes e_n$

converges for each $x \in \mathcal{H}$ in the Hilbert space tensor product $\mathcal{H} \otimes K$. In fact the series is Cauchy, hence convergent, as for each $M, N \in \mathbb{N}$ with $M < N - 1$,

$$\left\| \sum_{n=M+1}^{N} L_n x \otimes e_n \right\|^2 = \sum_{n=M+1}^{N} \langle L_n^* L_n x, x \rangle$$

$$\leq \left\| \sum_{n=M+1}^{N} L_n^* L_n x \right\| \|x\|$$

$$\to 0 \text{ as } M, N \to \infty.$$

Now consider the linear mapping L from $\mathcal{H}$ to $\mathcal{H} \otimes K$ given for each $x \in \mathcal{H}$ by

$$Lx = \sum_{n \in \mathbb{N}} L_n x \otimes e_n.$$

To show that it is bounded, by a similar calculation to that above, we have, by boundedness of the operator $\sum_{n=1}^{\infty} L_n^* L_n$, that there exists $C \geq 0$ such that

$$\|Lx\|^2 = \left\| \sum_{n=1}^{\infty} L_n x \otimes e_n \right\|^2 \leq \left\| \sum_{n=1}^{\infty} L_n^* L_n x \right\| \|x\| \leq C \|x\|^2.$$

Next, for $M, N \in \mathbb{N}$, $M < N - 1$, we let $Q_{M,N}$ be the orthogonal projection in K onto the linear span of $\{e_{M+1}, \dots, e_N\}$. If $x, y \in H$, a straightforward calculation yields

$$\langle (I \otimes Q_{M,N})(X \otimes I) Lx, Ly \rangle = \left\langle \sum_{n=M+1}^{N} L_n^* X L_n x, y \right\rangle.$$

Hence

$$\|L^*(I \otimes Q_{M,N})(X \otimes I) Lx\| = \sup_{\|y\|=1} |\langle (I \otimes Q_{M,N})(X \otimes I) Lx, Ly \rangle|$$

$$= \sup_{\|y\|=1} \left| \left\langle \sum_{n=M+1}^{N} L_n^* X L_n x, y \right\rangle \right|$$

$$= \left\| \sum_{n=M+1}^{N} L_n^* X L_n x \right\|.$$

Now by our earlier discussion, since $(I \otimes Q_{M,N})Lx = \sum_{n=M+1}^{N} L_n x \otimes e_n$, we have

$$\|L^*(I \otimes Q_{M,N})(X \otimes I)Lx\| = \|L^*(X \otimes I)(I \otimes Q_{M,N})Lx\|$$
$$\leq \|L^*\| . \|X\| . \|(I \otimes Q_{M,N})Lx\| \to 0,$$

as $M, N \to \infty$, and the result follows. $\qquad\square$

Observe that we recapture the Heisenberg evolution of reversible quantum theory if $L_n = 0$ for all $n \in \mathbb{N}$ in (4.4.8). The norm continuity assumption in Theorem 4.4.12 is somewhat artificial for bona fide physical applications. In the last part of this section, we try to give some non-rigorous, but hopefully useful insight into how the non-Hamiltonian terms in the Lindblad generator (4.4.12) arise. Let $B = (B(t), t \geq 0)$ be a standard Brownian motion defined on some probability space $(\Omega, \mathcal{F}, P)$, and let L be a bounded self-adjoint operator acting on $\mathcal{H}$. Consider the one-parameter unitary group $(U(t), t \in \mathbb{R})$ given for each $t \in \mathbb{R}$ by

$$U(t) = e^{itL} = \int_{\mathbb{R}} e^{it\lambda} E(d\lambda),$$

where E is the resolution of the identity associated to L via the spectral theorem. For $t > 0$, we employ the Brownian motion to randomise our unitary group and consider the unitary operator-valued stochastic process $U_B = (U_B(t), t \geq 0)$ defined for each $\omega \in \Omega$ by

$$U_B(t)(\omega) = e^{iB(t)(\omega)L} = \left(\int_{\mathbb{R}} e^{iB(t)\lambda} E(d\lambda) \right)(\omega). \qquad (4.4.9)$$

By Itô's formula (see section 3.1.2), we may write for each $t \geq 0, \lambda \in \mathbb{R}$,

$$e^{iB(t)\lambda} = 1 + i\lambda \int_0^t e^{iB(s)\lambda} dB(s) - \lambda^2 \int_0^t e^{iB(s)\lambda} ds,$$

and combining this with (4.4.9), we find that

$$U_B(t) = I + iL \int_0^t U_B(s) dB(s) - L^2 \int_0^t U_B(s) ds,$$

which we can write as an *operator-valued stochastic differential equation*:

$$dU_B(t) = iU_B(t)L\,dB(t) - U_B(t)L^2 dt. \qquad (4.4.10)$$

Note that L commutes with $U_B(t)$ for all $t \geq 0$, so it is a matter of convenience as to where we write it in this equation.

Taking formal adjoints in (4.4.10), we have

$$dU_B(t)^* = -iLU_B(t)^* dB(t) - L^2 U_B(t)^* dt. \qquad (4.4.11)$$

Now consider the family $(j_t^B, t \geq 0)$ of random automorphisms of $\mathcal{L}(H)$ defined by $j^B(t)(X) = U^B(t)XU^B(t)^*$ for each $X \in \mathcal{L}(H), t \geq 0$. Let us formally apply Itô's product formula (see, e.g., section 4.4.3 of Applebaum [6] pp. 257–8, particularly Example 4.4.15 therein). Then we get the following:

$$dj^B(t)(X) = dU^B(t)XU^B(t)^* + U^B(t)XdU^B(t)^* + dU^B(t)XdU^B(t)^*,$$

where the "Itô correction term" $dU^B(t)XdU^B(t)^*$ is formally obtained by replacing $dB(t)^2$ with $2dt$, and all other products of differentials with 0. Collecting terms, we find that

$$dj^B(t)(X) = j^B(t)(iLX - iXL)dB(t)$$
$$+ - j^B(t)(L^2X - 2LXL + XL^2)dt. \qquad (4.4.12)$$

We claim, but do not prove, that the prescription $T_t(X) = \mathbb{E}(j^B(t)(X))$ defines a norm-continuous quantum dynamical semigroup on $\mathcal{L}(H)$. We can calculate the generator using (4.4.12), noting that the stochastic integral will have mean zero, and so for all $t \geq 0$

$$T_t(X) - X = \int_0^t T_s A(X)ds,$$

where $A(X) = -(L^2X - 2LXL + XL^2)$ is clearly of the form (4.4.8). To obtain the general form of (4.4.8) by probabilistic arguments, one can develop a more sophisticated argument than the above using Brownian motion, as in Alicki and Fannes [5]. Alternatively, one can use quantum noise; indeed, from that perspective, to see how we can obtain terms involving non–self-adjoint L (so that both L and L^* play a role), we realise Brownian motion as an operator in a suitable Fock space and then split it into (mutually adjoint) creation and annihilation parts. To see this in full detail, with rigorous mathematical arguments, see Parthasarathy [73] and also Lindsay [63].

4.5 Exercises for Chapter 4

1. Let T be a symmetric linear operator in a Hilbert space.
    (a) Show that if T has an eigenvalue, then it is real valued.
    (b) Show that if T is self-adjoint then $\text{Ker}(T \pm iI) = \{0\}$. (Hint: Use a proof by contradiction).
    (c) If $\lambda \in \mathbb{C} \setminus \mathbb{R}$, show that the *Cayley transform*,

    $$U = (T - \bar{\lambda}I)(T - \lambda I)^{-1},$$

    is an isometry from $\text{Ran}(T - \lambda I)$ to $\text{Ran}(T - \bar{\lambda}I)$.

2. Establish the *basic criterion of self-adjointness*, i.e., that the following are equivalent for a symmetric linear operator T in a Hilbert space:
   (a) T is self-adjoint.
   (b) T is closed and $\text{Ker}(T^* \pm iI) = \{0\}$.
   (c) $\text{Ran}(T \pm iI) = H$.
   [Hint: Prove (a) $\Rightarrow$ (b) $\Rightarrow$ (c) $\Rightarrow$ (a). The hardest implication is the second of these. To establish this, use a proof by contradiction to show that (b) implies that $\text{Ran}(T \pm iI)$ is dense in H; then show that this set is also closed. For (c) $\Rightarrow$ (a), you will need the easily established fact that for any densely defined linear operator S in H, $\text{Ran}(S + iI) = H$ implies that $\text{Ker}(S^* - iI) = \{0\}$.]

3. If T is a self-adjoint operator in a Hilbert space H, deduce that its Cayley transform is unitary for $\lambda = \pm i$. What can you say about more general λ?

4. Use the closed graph theorem to prove the *Hellinger–Toeplitz theorem*: if T is a symmetric linear operator in a Hilbert space and $D_T = H$, then T is bounded, and so self-adjoint.

5. Let $f \in C_b(\mathbb{R}^d)$ and consider the associated *multiplication operator* M_f on $L^2(\mathbb{R}^d)$ defined by $M_f(\psi) = f\psi$, for $\psi \in L^2(\mathbb{R}^d)$. Is M_f symmetric, bounded, self-adjoint? Under what conditions does M_f generate a self-adjoint contraction semigroup?

6. Explain why the relativistic Schrödinger operator of Exercise 3.7 is self-adjoint.

7. Confirm that the deterministic dynamics given by the Heisenberg evolution is completely positive.

5

Compact and Trace Class Semigroups

5.1 Compact Semigroups

A bounded linear operator X acting in a Banach space E is said to be *compact* if for every bounded sequence (f_n) in E, the sequence $(Xf_n, n \in \mathbb{N})$ has a convergent subsequence. Let $\mathcal{K}(E)$ denote the space of all compact operators in E. It is easy to check that $\mathcal{K}(E)$ is a linear space. It is also not hard to see that $\mathcal{K}(E)$ is a two-sided ideal of $\mathcal{L}(E)$, i.e., for each $X \in \mathcal{K}(E)$ and $Y_1, Y_2 \in \mathcal{L}(E)$, we have $XY_1 \in \mathcal{K}(E)$ and $Y_2 X \in \mathcal{K}(E)$ (see Problem 5.1).

Suppose that $(T_t, t \geq 0)$ is a C_0-semigroup in E. If T_{t_0} is compact for some $t_0 \geq 0$, then it is compact for all $t > t_0$ (see Problem 5.2 (a)). If $\dim(E) = \infty$, then we must take $t_0 > 0$, as I is never compact.

We are particularly interested in the case where E is a Hilbert space (which, as usual, we denote by H) and X is self-adjoint as well as being compact. The following is a very well-known result (see, e.g., Reed and Simon [76] Theorem VI.16, p. 203 for a proof).

Theorem 5.1.1 (The Hilbert–Schmidt Theorem) *If X is a compact self-adjoint operator in a Hilbert space H, then there exists a complete orthonormal basis of eigenvectors $(e_n, n \in \mathbb{N})$ for H. The corresponding eigenvalues $(c_n, n \in \mathbb{N})$ are monotonically decreasing to zero as $n \to \infty$.*

Hence if X is compact and self-adjoint and $\psi \in H$, we may write down the Fourier expansion of the vector $X\psi$ with respect to the complete orthonormal basis $(e_n, n \in \mathbb{N})$ to obtain

$$X\psi = \sum_{n \in \mathbb{N}} c_n \langle \psi, e_n \rangle e_n. \tag{5.1.1}$$

We now explore the implications of the Hilbert–Schmidt theorem for semigroups.

Theorem 5.1.2 *If* $(T_t, t \geq 0)$ *is a* C_0-*semigroup in* H *that is both compact and self-adjoint for all* $t > 0$, *then there exists a sequence of positive numbers* $(\lambda_n, n \in \mathbb{N})$ *that diverges monotonically to infinity as* $n \to \infty$, *such that for all* $\psi \in H$,

$$T_t \psi = \sum_{n \in \mathbb{N}} e^{-t\lambda_n} \langle \psi, e_n \rangle e_n, \qquad (5.1.2)$$

where $(e_n, n \in \mathbb{N})$ *is as in Theorem 5.1.1.*

Proof. If T_t is compact, then by (5.1.1), for all $t > 0$, we have

$$T_t \psi = \sum_{n \in \mathbb{N}} c_n(t) \langle \psi, e_n \rangle e_n,$$

so we need only prove that $c_n(t) = e^{-\lambda_n t}$ for all $n \in \mathbb{N}, t > 0$. Since $T_t e_n = c_n(t) e_n$, we have for all $s, t > 0$,

$$c_n(s + t) e_n = T_{s+t} e_n = T_s T_t e_n = c_n(s) c_n(t) e_n,$$

and taking inner products of both sides of the last identity with e_n yields the functional equation $c_n(s+t) = c_n(s) c_n(t)$. Since (5.1.1) also holds for $T_0 = I$, we may extend the domain of the function c_n to $[0, \infty)$ with $c_0 = 1$. By the Parseval identity,

$$\begin{aligned}
\|(T_{s+t} - T_t)\psi\|^2 &= \left\| \sum_{n \in \mathbb{N}} (c_n(s + t) - c_n(t)) \langle \psi, e_n \rangle e_n \right\|^2 \\
&= \sum_{n \in \mathbb{N}} (c_n(s + t) - c_n(t))^2 \langle \psi, e_n \rangle^2 \\
&\to 0 \text{ as } s \to 0.
\end{aligned}$$

It follows by dominated convergence that $t \to c_n(t)$ is right continuous. The key estimate we need to implement this is that fact that for sufficiently large T, using (1.2.4),

$$\begin{aligned}
|c_n(s + t) - c_n(t)| &= |\langle (T_{s+t} - T_t) e_n, e_n \rangle| \\
&\leq \|T_{s+t}\| + \|T_t\| \leq 2Me^{aT}.
\end{aligned}$$

A similar argument establishes left continuity. But (see, e.g., Bingham et al. [14], pp. 4–5) the only continuous solutions of the functional equation that have the required limiting behaviour are $c_n(t) = e^{-\lambda_n t}$. Since for all $t > 0$ the sequence $(c_n(t))$ is monotonic decreasing to zero by Theorem 5.1.1, it follows that (λ_n) is monotonic increasing to infinity. $\square$

Note that if $e_n \in D_A$ and $Ae_n = -\lambda_n e_n$ for each $n \in \mathbb{N}$, then (5.1.2) follows by functional calculus.

For an example, consider our standard parabolic PDE (2.3.4), but with $b = 0$, then for all $f \in C_c^2(\Omega)$, $x \in \Omega$, we have

$$Af(x) = \sum_{i=1}^{n} \partial_i (a_{ij}(x)\partial_j) f(x) - c(x)f(x).$$

It is easy to see that A is symmetric and that $-A$ is positive. Indeed, using integration by parts we have

$$\langle Af, f \rangle = -\sum_{i,j=1}^{d} \int_{\Omega} a_{ij}(x)\partial_i f(x)\partial_j f(x)dx - \int_{\omega} c(x)|f(x)|^2 dx \leq 0.$$

Then the Friedrichs extension $-A_F$ of $-A$ (cf. Theorem 4.1.5) is a positive self-adjoint operator, and the unique solution of the parabolic PDE is implemented by the semigroup $T_t = e^{tA_F}$, for $t \geq 0$, acting in $L^2(\Omega) := L^2(\Omega; \mathbb{R})$. Now we know from the work of section 2.3 that this unique weak solution $u(t, \cdot)$ lies in $H_0^1(\Omega)$. Let $Y_t \colon L^2(\Omega) \to H_0^1(\Omega)$ be the mapping that takes each initial condition $f \in L^2(\Omega)$ to its solution in $H_0^1(\Omega)$. Then we have for all $t > 0$, $T_t = Y_t \Theta$, where Θ is the embedding of $H_0^1(\Omega)$ into $L^2(\Omega)$. But Y_t is a bounded operator and Θ is compact by the Rellich–Kondrachov theorem (Theorem C.0.2). Hence T_t is compact and self-adjoint. The semigroup $(T_t, t \geq 0)$ is called a *symmetric diffusion semigroup*.

5.2 Trace Class Semigroups

5.2.1 Hilbert–Schmidt and Trace Class Operators

We sketch some well-known ideas in this introductory subsection. For more details (see, e.g., Chapter VI, section 6 in Reed and Simon [76], pp. 206–13). The problems at the end of the chapter will also explore some of this territory. Let H be a complex separable Hilbert space. We say that a bounded linear operator T acting on H is Hilbert–Schmidt if $\sum_{n=1}^{\infty} ||Te_n||^2 < \infty$ for some (and hence all) complete orthonormal basis $(e_n, n \in \mathbb{N})$ in H. Let $\mathcal{I}_2(H)$ denote the linear space of all Hilbert–Schmidt operators on H. We have the following basic facts:

1. $\mathcal{I}_2(H)$ is a (two-sided) *-ideal of $\mathcal{L}(H)$.
2. $\mathcal{I}_2(H) \subseteq \mathcal{K}(H)$.

There is a very precise description of Hilbert–Schmidt operators acting in L^2 spaces, which is presented in the next theorem.

Theorem 5.2.3 *Let (S, Σ, μ) be a measure space wherein the σ-algebra Σ is countably generated. Let $H = L^2(S, \Sigma, \mu; \mathbb{C})$. Then $T \in \mathcal{I}_2(H)$ if and only if for all $f \in H, x \in S$,*

$$(Tf)(x) = \int_S f(y)k(x, y)\mu(dy), \qquad (5.2.3)$$

where the kernel $k \in L^2(S \times S, \Sigma \otimes \Sigma, \mu \times \mu; \mathbb{C})$.

Proof. We only sketch sufficiency here. Since Σ is countably generated, H is separable. Let $(e_n, n \in \mathbb{N})$ be a complete orthonormal basis for H. For each $x \in S$, define $k_x(y) = k(x, y)$. Then, for all $x \in S \setminus N$, where $\mu(N) = 0$, $k_x \in L^2(S)$ and for all $n \in \mathbb{N}$,

$$(Te_n)(x) = \int_S k(x, y)e_n(y)\mu(dy) = \langle k_x, \overline{e_n} \rangle.$$

Then we have, by Fubini's theorem and the Pythagorean theorem,

$$\sum_{n=1}^{\infty} ||Te_n||^2 = \sum_{n=1}^{\infty} ||\langle k., \overline{e_n} \rangle||^2$$

$$= \int_S \sum_{n=1}^{\infty} |\langle k_x, \overline{e_n} \rangle|^2 \mu(dx)$$

$$= \int_S ||k_x||^2 \mu(dx) = \int_S \int_S |k(x, y)|^2 \mu(dx)\mu(dy) < \infty,$$

and so T is Hilbert–Schmidt, as required. $\qquad \square$

Any bounded linear operator T in H which takes the form (5.2.3) is called an *integral operator* with kernel k. It is straightforward to check that if T is an integral operator, then so is T^* with

$$T^*f(x) = \int_S f(y)\overline{k(y, x)}\mu(dx),$$

for all $f \in H, x \in S$. Furthermore T is self-adjoint if and only if $k(x, y) = \overline{k(y, x)}$ for all $(x, y) \in (S \times S) \setminus N$ with $N \in \Sigma \otimes \Sigma$ such that $(\mu \times \mu)(N) = 0$.

Next we turn our attention to trace class operators. If $T \in \mathcal{L}(H)$ the operator T^*T is positive and self-adjoint. Hence by spectral theory, it has a positive self-adjoint square root which we denote as $|T| = (T^*T)^{\frac{1}{2}}$. In particular, if T is positive self-adjoint, then $|T| = T$. We say that T is *trace class* if

$$\text{tr}(|T|) := \sum_{n=1}^{\infty} \langle |T|e_n, e_n \rangle < \infty,$$

for some (and hence all) complete orthonormal basis $(e_n, n \in \mathbb{N})$ of H. The linear space of all trace class operators on H is denoted $\mathcal{I}_1(H)$.

1. $\mathcal{I}_1(H)$ is a (two-sided) *–ideal of $\mathcal{L}(H)$.
2. $\mathcal{I}_1(H) \subseteq \mathcal{K}(H)$.

Proposition 5.2.4 $\mathcal{I}_1(H) \subseteq \mathcal{I}_2(H)$.

Proof. If $T \in \mathcal{I}_1(H)$, then using the representation (5.1.1) we have. $\text{tr}(|T|) = \sum_{n=1}^{\infty} |c_n| < \infty$. By the same representation, we have

$$\sum_{n=1}^{\infty} \|T e_n\|^2 = \sum_{n=1}^{\infty} |c_n|^2 \leq \left(\sum_{n=1}^{\infty} |c_n| \right)^2 < \infty,$$

and so $T \in \mathcal{I}_2(H)$, as required. $\qquad\square$

It is an interesting fact that a bounded linear operator in H is trace class if and only if it can be written as the product of two Hilbert–Schmidt operators. We will only prove a very special case of this result.

Proposition 5.2.5 *If A, B are positive self-adjoint commuting Hilbert–Schmidt operators, then AB is trace class.*

Proof. Clearly AB is positive and self-adjoint and

$$\begin{aligned}
\text{tr(AB)} &= \sum_{n=1}^{\infty} \langle AB e_n, e_n \rangle \\
&= \sum_{n=1}^{\infty} \langle B e_n, A e_n \rangle \\
&\leq \sum_{n=1}^{\infty} \|B e_n\| . \|A e_n\| \\
&\leq \left(\sum_{n=1}^{\infty} \|B e_n\|^2 \right)^{\frac{1}{2}} \left(\sum_{n=1}^{\infty} \|A e_n\|^2 \right)^{\frac{1}{2}} < \infty,
\end{aligned}$$

and the result follows. $\qquad\square$

If T is a trace class operator, it can be shown that its trace $\text{tr(T)} := \sum_{n=1}^{\infty} \langle T e_n, e_n \rangle$ is finite (in fact the series converges absolutely; see Theorem VI.24 in Reed and Simon [76], p. 211). If T is also positive and self-adjoint, then $\text{tr(T)} = \sum_{n=1}^{\infty} \lambda_n$, where $\{\lambda_n, n \in \mathbb{N}\}$ are the eigenvalues of T.

Again we will only sketch part of the proof of the following beautiful theorem.

Theorem 5.2.6 (Mercer's Theorem) *Let* (S, Σ, μ) *be a measure space wherein the σ-algebra Σ is countably generated. Let $H = L^2(S, \Sigma, \mu; \mathbb{C})$, and for all $f \in H, x \in S$ define the Hilbert–Schmidt operator*

$$(Af)(x) = \int_S f(y)k(x, y)\mu(dy),$$

where the kernel $k \in L^2(S \times S, \Sigma \otimes \Sigma, \mu \times \mu; \mathbb{C})$. Assume that k is continuous and A is positive and self-adjoint. Then A is trace class if and only if $\int_S |k(x, x)|\mu(dx) < \infty$, in which case

$$\mathrm{tr}(A) = \int_S k(x, x)\mu(dx).$$

Proof. We will only prove the last part here. For a full proof, see Davies [27], pp. 156–7. Define k_x as in the proof of Theorem 5.2.3. Let $(e_n, n \in \mathbb{N})$ be a complete orthonormal basis of eigenvectors for A so that $Ae_n = \lambda_n e_n$, for all $n \in \mathbb{N}$. Then, for all $x \in S$,

$$k_x = \sum_{n=1}^{\infty} \langle k_x, e_n \rangle e_n$$

$$= \sum_{n=1}^{\infty} \overline{Ae_n(x)} e_n$$

$$= \sum_{n=1}^{\infty} \lambda_n \overline{e_n(x)} e_n.$$

Then,

$$\int_S k(x, x)\mu(dx) = \int_S k_x(x)\mu(dx)$$

$$= \sum_{n=1}^{\infty} \lambda_n \int_S |e_n(x)|^2 \mu(dx)$$

$$= \sum_{n=1}^{\infty} \lambda_n = \mathrm{tr}(A). \qquad \square$$

Within the last proof, we established that for all $x \in S$,

$$k_x = \sum_{n=1}^{\infty} \lambda_n \overline{e_n(x)} e_n,$$

and so if $y \in S$, we have

$$k(x, y) = \sum_{n=1}^{\infty} \lambda_n \overline{e_n(x)} e_n(y), \qquad (5.2.4)$$

whenever the series converges pointwise everywhere, e.g., if the eigenfunctions are uniformly bounded. This is very useful in the theory of reproducing kernel Hilbert spaces.

5.2.2 Trace Class Semigroups

We say that a semigroup $(T_t, t \geq 0)$ is *trace class* if $T_t \in \mathcal{I}_1(H)$ for all $t > 0$ and *Hilbert–Schmidt* if $T_t \in \mathcal{I}_2(H)$ for all $t > 0$.

Proposition 5.2.7 *A C_0-semigroup $(T_t, t \geq 0)$ is trace class if and only if it is Hilbert–Schmidt.*

Proof. If T_t is trace class, then it is Hilbert–Schmidt by Proposition 5.2.4. Conversely, suppose the semigroup is Hilbert–Schmidt; then since for all $t > 0, T_t = T_{t/2} T_{t/2}$, we see that T_t is trace class by the remark preceding Proposition 5.2.5. $\qquad\square$

We will only touch on the subject of trace class semigroups here. A full monograph treatment of this topic, where they are called "Gibbs semigroups", is Zagrebnov [103].

Now suppose that $(T_t, t \geq 0)$ is a trace class semigroup on $L^2(S, \Sigma, \mu; \mathbb{C})$. Then by Proposition 5.2.7 and Theorem 5.2.3, we can write

$$T_t f(x) = \int_S f(y) k_t(x, y) \mu(dy),$$

for all $t > 0, f \in L^2(S), x \in S$. Let us assume further that $(T_t, t \geq 0)$ is self-adjoint. Then by Theorem 5.1.2, there exists a complete orthonormal basis $(e_n, n \in \mathbb{N})$ for $L^2(S)$ and a sequence (λ_n) of positive real numbers that diverges monotonically to infinity as $n \to \infty$, so that $T_t e_n = e^{-t\lambda_n} e_n$ for all $t > 0, n \in \mathbb{N}$.

Proposition 5.2.8 *Suppose that $(T_t, t \geq 0)$, $(e_n, n \in \mathbb{N})$ and $(\lambda_n, n \in \mathbb{N})$ are as above. If e_n is continuous for all $n \in \mathbb{N}$ and*

$$\sum_{n=1}^{\infty} e^{-t\lambda_n} \|e_n\|_{\infty}^2 < \infty$$

for all $t > 0$, then k is continuous on $(0, \infty) \times S \times S$.

Proof. This is based on Theorem 7.2.9 in Davies ([27], p. 195). By (5.2.4) and the condition in the statement of the theorem, we have

$$k_t(x, y) = \sum_{n=1}^{\infty} e^{-t\lambda_n} \overline{e_n(x)} e_n(y),$$

and this is a uniformly convergent series of continuous functions (on compact intervals of $(0, \infty)$). The result follows. □

We also note the useful fact that a self-adjoint semigroup $(T_t, t \geq 0)$ such that T_t has a complete orthonormal basis of eigenvectors, with corresponding eigenvalues $\{e^{-t\lambda_n}, n \in \mathbb{N}\}$ for all $t > 0$, is trace class if and only if for all $t > 0$,

$$\text{tr}(T_t) = \sum_{n=1}^{\infty} e^{-t\lambda_n} < \infty. \tag{5.2.5}$$

This is easily deduced from (5.1.2) and the definition of the trace.

5.2.3 Convolution Semigroups on the Circle

As an example of an interesting class of trace class semigroups, we consider convolution semigroups on the circle. Let $S^1 := \mathbb{R}/2\pi\mathbb{Z}$ be the circle of circumference 2π, and $\natural: \mathbb{R} \to S^1$ be the usual quotient map which takes each $x \in \mathbb{R}$ to the coset $[x] := \{x + 2n\pi; n \in \mathbb{Z}\}$. So $\natural$ is a surjective homomorphism of abelian groups, where the group law in S^1 is given by $[x] + [y] = x + y \mod 2\pi$ for each $x, y \in \mathbb{R}$. The collection of all minor arcs (with end points omitted) forms a basis for the topology of S^1, with respect to which $\natural$ is both open and continuous. Every continuous function f on S^1 pulls back to a continuous periodic function $f^\natural$ on $\mathbb{R}$, where $f^\natural := f \circ \natural$. Given a Borel measure ρ on $\mathbb{R}$, its pushforward $\widetilde{\rho} := \rho \circ \natural^{-1}$ is a Borel measure on S^1. In particular, any positive multiple of Lebesgue measure on $\mathbb{R}$ gives rise in this way to a (finite) Haar measure m on the circle. We normalise so that $m(S_1) = 1$, and for all $f \in C(S^1)$, we write $\int_{S^1} f([x])m(d[x]) = \int_{S^1} f([x])d[x]$, where $d[x] = dx/2\pi$.

Theorem 5.2.9 *If $(\mu_t, t \geq 0)$ is a convolution semigroup of probability measures on $\mathbb{R}$, then $(\widetilde{\mu}_t, t \geq 0)$ is a convolution semigroup of probability measures on S^1.*

Proof. For all $f \in C(S^1), s, t \geq 0$,

$$\int_{S^1} f([x])\widetilde{\mu_{s+t}}(d[x]) = \int_{\mathbb{R}} f^\natural(x)\mu_{s+t}(dx)$$

$$= \int_{\mathbb{R}} \int_{\mathbb{R}} f^\natural(x + y)\mu_s(dx)\mu_t(dy)$$

$$= \int_{\mathbb{R}} \int_{\mathbb{R}} f(\natural(x) + \natural(y)) \mu_s(dx) \mu_t(dy)$$

$$= \int_{S^1} \int_{S^1} f([x] + [y]) \widetilde{\mu}_s(dx) \widetilde{\mu}_t(dy),$$

so that $\widetilde{\mu_{s+t}} = \widetilde{\mu}_s * \widetilde{\mu}_t$.

Clearly $\widetilde{\mu}_0 = \delta_0 \circ \natural^{-1} = \delta_{[0]}$. Finally, for all $f \in C(S^1)$,

$$\lim_{t \to 0} \int_{S^1} f([x]) \widetilde{\mu}_t(d[x]) = \lim_{t \to 0} \int_{S^1} f(\natural(x)) \mu_t(dx) = f(\natural(0)) = f([0]),$$

so we have the required weak continuity, and the proof is complete. $\qquad\square$

Just as on the real line, we may define semigroups $(\widetilde{T}_t, t \geq 0)$ on $C(S^1)$ and on $L^p(S^1)$ by the prescription

$$\widetilde{T}_t f(x) = \int_{S^1} f([x] + [y]) \widetilde{\mu}_t(d[x]),$$

for all $t \geq 0$, $x \in \mathbb{R}$, $f \in C(S^1)$ (say). Then it is not difficult to verify that

$$\widetilde{T}_t f \circ \natural = T_t(f \circ \natural),$$

where $(T_t, t \geq 0)$ is the C_0-semigroup acting in $L^2(\mathbb{R})$ that is induced by $(\mu_t, t \geq 0)$. If for all $t \geq 0$, T_t is self-adjoint in $L^2(\mathbb{R})$, then so is $\widetilde{T}_t$ in $L^2(S^1)$ (see Problem 5.7). We compute the spectrum of $\widetilde{T}_t$ with respect to the complete orthonormal basis $(e_n, n \in \mathbb{Z})$ of $L^2(S^1)$, where $e_n([x]) := e^{inx}$ for each $x \in \mathbb{R}$, $n \in \mathbb{Z}$. For each $t \geq 0$, we have

$$\widetilde{T}_t e_n([x]) = \int_{S^1} e_n([x] + [y]) \widetilde{\mu}_t(d[y])$$

$$= e_n([x]) \int_{\mathbb{R}} e^{iny} \mu_t(dy)$$

$$= e^{t\eta(n)} e_n([x]),$$

so that T_t has eigenvalues $\{e^{t\eta(n)}, n \in \mathbb{Z}\}$, where η is given by the Lévy–Khintchine formula (3.4.22). From now on we assume that $(T_t, t \geq 0)$ is self-adjoint. Equivalently, by Theorem 4.2.8 (iv) we have

$$\eta(n) = -an^2 + \int_{\mathbb{R}} (\cos(nx) - 1) \nu(dx),$$

where $a \geq 0$ and the Lévy measure ν is symmetric. By (5.2.5) we see that $\widetilde{T}_t$ is trace class for all $t > 0$ if and only if

$$\sum_{n \in \mathbb{Z}} e^{t\eta(n)} < \infty. \tag{5.2.6}$$

Since $\|e_n\|_\infty = 1$ for all $n \in \mathbb{Z}$, it follows by Proposition 5.2.8 that for all $t > 0$, $\widetilde{T}_t$ is an integral operator with a (uniformly) continuous kernel k_t if (5.2.6) holds. If $a \neq 0$, this condition is always satisfied as

$$\sum_{n \in \mathbb{Z}} e^{t\eta(n)} \leq \sum_{n \in \mathbb{Z}} e^{-atn^2} < \infty.$$

We assume henceforth that (5.2.6) holds. We then have for all $f \in C(S^1)$,

$$\widetilde{T}_t f([0]) = \int_{S^1} f([y]) k_t([0], [y]) d[y] = \int_{S^1} f([y]) \widetilde{\mu}_t(d[y]),$$

and so $\widetilde{\mu}_t$ is absolutely continuous with respect to Lebesgue measure on S^1, with density $h_t(\cdot) = k_t([0], \cdot)$. For all $x \in \mathbb{R}$, we then have

$$\widetilde{T}_t(x) = \int_{S^1} f([x] + [y]) h_t([y]) d[y] = \int_{S^1} f([y]) h_t([y] - [x]) d[y],$$

so that $k_t([x], [y]) = h_t([y] - [x])$, for all $t > 0, x, y \in \mathbb{R}$.

We now apply Mercer's theorem (Theorem 5.2.6). Since for all $t > 0$,

$$\int_{S^1} k_t([x], [x]) d[x] = h_t([0]) m(S^1) = h_t([0]),$$

we have the elegant trace formula:

$$\mathrm{tr}(\widetilde{T}_t) = h_t([0]) = \sum_{n \in \mathbb{Z}} e^{t\eta(n)}. \tag{5.2.7}$$

These ideas extend naturally to the d-torus, which is just the direct product of d copies of S^1. It can be generalised beyond that context to compact (not necessarily abelian) Lie groups. For details, see Chapter 5 of Applebaum [7].

As an interesting special case of (5.2.7), consider the rescaled Brownian motion on $\mathbb{R}$, $B = (B(t), t \geq 0)$, where for each $t \geq 0$, $B(t)$ has variance σt, with $\sigma > 0$. This corresponds to a convolution semigroup $(\mu_t, t \geq 0)$, having characteristics $(0, \sigma^2/2, 0)$. For each $t > 0$, μ_t has density f_t, where for each $x \in \mathbb{R}$,

$$f_t(x) = \frac{1}{\sigma\sqrt{2\pi t}} e^{-\frac{|x|^2}{2\sigma^2 t}}.$$

Then $\widetilde{\mu}_t$ has density h_t, with respect to *normalised* Lebesgue measure, which is given by scaling and periodisation of f_t, i.e., for all $x \in \mathbb{R}$,

$$h_t([x]) = 2\pi \sum_{n \in \mathbb{Z}} f_t(x + 2\pi n) = \frac{\sqrt{2\pi}}{\sigma t} \sum_{n \in \mathbb{Z}} e^{\frac{-|x + 2\pi n|^2}{2\sigma^2 t}}.$$

We have $\eta(n) = -\sigma^2 n^2/2$, for all $n \in \mathbb{N}$. Now choose $\sigma = \sqrt{2\pi}$, and define the classical *theta function* $\theta(t)$, for $t > 0$, by

$$\theta(t) := \sum_{n \in \mathbb{Z}} e^{-\pi n^2 t}.$$

With our given choice of η and σ, we have $\theta(t) = \sum_{n \in \mathbb{Z}} e^{t\eta(n)}$, and in this case (5.2.7) is easily seen to be nothing but the celebrated functional equation for the theta function:

$$\theta(t) = \frac{1}{\sqrt{t}}\theta(1/t),$$

for all $t > 0$. This is usually derived from the Poisson summation formula (see, e.g., section 3.1 of Stein and Shakarchi [94], p. 155). For more about the relationship between (5.2.7) and the Poisson summation formula, see Applebaum [8].

5.2.4 Quantum Theory Revisited

This part is not strictly about trace class semigroups, but it does involve both unitary groups and trace class operators, so what better place to put it? Let $\mathcal{H}$ be a complex separable Hilbert space, and $\mathcal{I}_1(\mathcal{H})$ be the space of all trace class operators on $\mathcal{H}$. A linear operator $\rho \in \mathcal{I}_1(\mathcal{H})$ is called a *density operator*[1] if it is positive and self-adjoint with $\mathrm{tr}(\rho) = 1$. Density operators describe *mixed states* in quantum theory. In basic quantum mechanics, we have already, in section 4.2, pointed out that a state of a system is described by a unit vector $\psi \in \mathcal{H}$. Each such ψ gives rise to a density operator which is simply the orthogonal projection P_ψ onto the linear span of ψ, so if $\phi \in \mathcal{H}$, then $P_\psi(\phi) = \langle \phi, \psi \rangle \psi$. Such projections are called *pure states*. Now suppose that we are in a state of ignorance. All we know about our system is that there are two pure states ψ_1 and ψ_2 which occur with probability c and $1 - c$, respectively. Then we have a *mixed state* that cannot be described by an orthogonal projection but is instead represented by the density operator

$$\rho = cP_{\psi_1} + (1 - c)P_{\psi_2}.$$

In general, if ρ is a density operator, *expectation* of the operator $X \in \mathcal{L}(\mathcal{H})$ is given by the linear mapping $\omega_\rho \colon \mathcal{L}(\mathcal{H}) \to \mathbb{C}$, where

$$\omega_\rho(X) = \mathrm{tr}(\rho X), \tag{5.2.8}$$

---

[1] Physicists sometimes use the term "density matrix".

for all $X \in \mathcal{L}(\mathcal{H})$. The mapping ω_ρ is called a *state* on the algebra $\mathcal{L}(\mathcal{H})$. Note that the ideal property of $\mathcal{I}_1(\mathcal{H})$ ensures that it is well-defined and takes finite values. Now $\mathcal{I}_1(\mathcal{H})$ is a complex Banach space under the trace norm $||T||_1 = \mathrm{tr}(|T|)$, and $\mathcal{L}(\mathcal{H})$ is in fact the dual space of $\mathcal{I}_1(\mathcal{H})$, with the duality given by the mapping $(T, X) \to \omega_T(X)$ for $T \in \mathcal{I}_1(\mathcal{H})$, $X \in \mathcal{L}(\mathcal{H})$ (see, e.g., Proposition 2.4.3 in Bratteli and Robinson [19], pp. 68–9). Then if $(T_t, t \geq 0)$ is a C_0-semigroup (or group) on $\mathcal{I}_1(\mathcal{H})$, then its adjoint $(T_t^*, t \geq 0)$ is a C_0-semigroup (or group) on $\mathcal{L}(\mathcal{H})$. In particular if $T_t(Y) = U(t)^*YU(t)$, for all $t \in \mathbb{R}$, $Y \in \mathcal{I}_1(\mathcal{H})$, where $(U(t), t \in \mathbb{R})$ is a one-parameter group, then by (5.2.8), $T_t^*(X) = U(t)XU(t)^*$, for all $t \in \mathbb{R}$, $X \in \mathcal{L}(\mathcal{H})$ is a Heisenberg evolution of observables. We use this implicitly in some computations that will appear very shortly.

Recall that in section 4.2, we discussed the Schrödinger evolution of pure states, and the Heisenberg evolution of operators. We may now generalise the former to mixed states. For each $t \in \mathbb{R}$ we define $\rho_t := U_t^* \rho U_t$, where $U_t = e^{itH}$ is as in section 4.2. It is easy to see that ρ_t is a density operator for all $t \in \mathbb{R}$. Then, since for all $X \in \mathcal{L}(\mathcal{H})$,

$$\mathrm{tr}(\rho_t X) = \mathrm{tr}(U_t^* \rho U_t X) = \mathrm{tr}(\rho U_t X U_t^*),$$

we have (arguing very informally),

$$\mathrm{tr}\left(\frac{d\rho_t}{dt} X\right) = \frac{d}{dt}\mathrm{tr}(\rho_t X)$$
$$= \mathrm{tr}\left(\rho \frac{dj_t(X)}{dt}\right)$$
$$= \mathrm{tr}(\rho \, ij_t([H, X]))$$
$$= i\,\mathrm{tr}(\rho_t(HX - XH))$$
$$= \mathrm{tr}(i[\rho_t, H]X).$$

This suggests that density operators evolve in accordance with the *quantum Liouville equation*

$$\frac{d\rho_t}{dt} = i[\rho_t, H],$$

and this is indeed the case.

An important example of a density operator arises in quantum statistical mechanics. Assume that the Hamiltonian operator H is positive and that $e^{-\beta H}$ is trace class where $\beta = 1/kT$ is the inverse temperature parameter, so T is the temperature of our system, and k is Boltzmann's constant. Then the *Gibbs density operator* is defined by $\rho = \frac{e^{-\beta H}}{\mathrm{tr}(e^{-\beta H})}$, and the quantity $\mathrm{tr}(e^{-\beta H})$ is the *quantum partition function.* For more about Gibbs states, and applications to Bose and Fermi gases, see, e.g., Bratteli and Robinson [20].

We can generalise this circle of ideas to the Lindblad equation (4.4.12). So if $(T_t, t \geq 0)$ is a norm-continuous quantum dynamical semigroup, we write[2]

$$\text{tr}(T_t^* \rho X) = \text{tr}(\rho T_t(X)).$$

By similar arguments to the derivation above of the quantum Liouville equation, we find that, writing $\rho_t = T_t^* \rho$, for $t \geq 0$, we have $\frac{d\rho_t}{dt} = \mathcal{L}^*(\rho_t)$, where

$$\mathcal{L}^* \rho = -i[H, \rho] + \frac{1}{2} \sum_{n \in \mathbb{N}} (L_n^* L_n \rho - 2 L_n \rho L_n^* + \rho L_n^* L_n).$$

We may try to generalise the discussion of mixed states a little further. A linear mapping $\omega \colon \mathcal{L}(\mathcal{H}) \to \mathbb{C}$ is called a *state* if $\omega(I) = 1$ and $\omega(X) \geq 0$ whenever $X \geq 0$, i.e., X is a positive self-adjoint operator. It is clear that every mixed state is a state. To see this, observe that if $X \geq 0$, then it has a positive self-adjoint square root $X^{1/2}$; indeed, using spectral decompositions, if $X = \int_{\mathbb{R}} \lambda E(d\lambda)$, then $X^{1/2} = \int_{\mathbb{R}} \lambda^{1/2} E(d\lambda)$. Then

$$\omega(X) = \text{tr}(\rho X) = \text{tr}(\rho^{1/2} X^{1/2} X^{1/2} \rho^{1/2}) = \text{tr}(B^* B) \geq 0,$$

where $B := X^{1/2} \rho^{1/2}$. A beautiful result, called *Gleason's Theorem*,[3] states that every state on $\mathcal{L}(\mathcal{H})$ is a mixed state, and so our generalisation appears to yield nothing new. However, if $\mathcal{L}(\mathcal{H})$ is replaced by a more general *-algebra, then Gleason's theorem breaks down, and there are examples of "non-normal" states that play important roles in quantum statistical mechanics.

5.3 Exercises for Chapter 5

1. Show that $\mathcal{K}(E)$ is a two-sided ideal of $\mathcal{L}(E)$.
2. Let $(T_t, t \geq 0)$ be a C_0-semigroup in E. Suppose that it is compact for some $t_0 > 0$.
   (a) Show that T_t is compact for all $t > t_0$.
   (b) Prove that the mapping $t \to T_t$ is norm continuous on (t_0, ∞). [Hint: Use the following fact that is proved in Engel and Nagel [32], p. 3, Lemma 1.2. If C is a compact set in $[0, \infty)$ and K is a compact set in

---

[2] There is some abuse of notation here, as $\mathcal{L}(H)$ is the dual of $\mathcal{I}_1(H)$.
[3] This is proved in, e.g., Parthasararthy [73] Chapter 1, section 8, pp. 31–40.

E, then the mapping $(t, \psi) \to T_t\psi$ from $C \times K \to E$ is uniformly continuous.]

3. (a) Let T be a positive self-adjoint bounded linear operator in a Hilbert space H. Prove that if $\sum_{n=1}^{\infty} \langle Te_n, e_n \rangle < \infty$ holds for some complete orthonormal basis (e_n) of H, then it holds for all such bases. [Hint: Write $T = S^2$.]

   (b) Show that if T is a trace class operator, then the expression $\sum_{n=1}^{\infty} \langle Te_n, e_n \rangle$ (which you may assume to be absolutely convergent) is independent of choice of complete orthonormal basis (e_n) of H.

4. Let H be a complex Hilbert space.

   (a) Show that any bounded linear operator in H may be written as a linear combination of two self-adjoint operators. (Hint: Think of "real" and "imaginary" parts.)

   (b) If A is a self-adjoint contraction in H, deduce that the operators $A \pm i(I - A^2)^{\frac{1}{2}}$ are unitary.

   (c) Use the results of (a) and (b) to show that any bounded linear operator in H may be written as a linear combination of four unitary operators.

   (d) Use the result of (c) to show that if A is trace class and B is bounded, then

$$\operatorname{tr}(AB) = \operatorname{tr}(BA).$$

5. Let $(R_\lambda, \lambda \in \rho(X))$ be the resolvent of a closed operator X acting in a Banach space E.

   (a) Show that if R_λ is compact for some $\lambda \in \rho(X)$, then it is compact for all $\lambda \in \rho(A)$.

   (b) If X is bounded and R_λ is compact for some/all $\lambda \in \rho(X)$, show that $\dim(E) < \infty$.

6. A C_0-semigroup $(T_t, t \geq 0)$ acting in a Banach space is said to have *compact resolvent* if its resolvent is compact, in the sense of Problem 5.5. Show that if $(T_t, t \geq 0)$ is a self-adjoint semigroup in a Hilbert space with compact resolvent, then its spectrum is discrete.

   (Warning: Having a compact resolvent tells us nothing about compactness of the semigroup, and vice versa.)

7. Check that $\tilde{T}_t$ is self-adjoint on $L^2(S)$ for all $t \geq 0$.

6

Perturbation Theory

This is a huge subject. But it is very easy to state the basic problem.

Main Problem of Perturbation Theory. Suppose that A is a linear operator in a Banach space E that generates a C_0-semigroup $(S_t, t \geq 0)$. Let B be another linear operator acting in E and assume that $A + B$ is at least densely defined and closeable. When does $A + B$ generate a C_0-semigroup?

If A is self-adjoint and B is symmetric, an auxiliary problem that is of some importance in quantum mechanics, is to find conditions for $A + B$ to at least be essentially self-adjoint.

6.1 Relatively Bounded and Bounded Perturbations

6.1.1 Contraction Semigroups

It seems obvious that our first attempt at solving the "main problem" should be in the case where B is bounded. But we are going to postpone this for a while and look at what at first appears to be a more general problem. In this section, we mainly follow Goldstein [40], pp. 38–40. We begin with a definition. Let A and B be linear operators in E with $D_A \subseteq D_B$. We say that B is *relatively bounded* with respect to A if there exists $0 \leq a \leq 1$ and $b \geq 0$ so that for all $f \in D_A$,

$$||Bf|| \leq a||Af|| + b||f||. \tag{6.1.1}$$

The A-*bound* of B, which we denote by a_0, is defined by

$$a_0 := \inf\{a \geq 0; \text{ there exists } b \geq 0 \text{ such that (6.1.1) holds}\}.$$

We see that a relatively bounded B is in fact bounded if we can take $a = 0$ in (6.1.1) and if D_B is dense in E.

Theorem 6.1.1 *Assume that A and B are linear operators in E so that:*

1. *A is the generator of a contraction semigroup in E.*
2. *B is dissipative.*
3. *B is relatively bounded with respect to A.*

Then, $A + B$ generates a contraction semigroup in E.

Proof. We observe that A is dissipative by the Lumer–Phillips theorem (Theorem 2.2.6), and then it can be verified that $A + B$ is dissipative on D_A. We omit the proof of this fact, pointing out only that it is easy when E is a Hilbert space.

Case I ($a < 1/2$).

By Theorem 2.2.7, it is sufficient to show that $\mathrm{Ran}(\lambda I - (A + B)) = E$ for some $\lambda > 0$. First note that the fact that $D_A \subseteq D_B$ implies that for all $\lambda > 0$,

$$\mathrm{Ran}(\lambda I - (A + B)) \supseteq \mathrm{Ran}((\lambda I - (A + B))R_\lambda(A)) = \mathrm{Ran}(I - BR_\lambda(A)),$$

since $R_\lambda(A)$ maps E to D_A. Consequently it is sufficient to show that $\|BR_\lambda(A)\| \leq 1$ for some $\lambda > 0$, for then $I - BR_\lambda(A)$ is invertible (with inverse given by a Neumann series), and the result will follow.

But for all $f \in E$, by (6.1.1), Lemma 2.1.1 and the fact that $\|R_\lambda(A)\| \leq 1/\lambda$, we have

$$\|BR_\lambda(A)f\| \leq a\|AR_\lambda(A)f\| + b\|R_\lambda(A)f\|$$
$$= a\|(\lambda R_\lambda(A) - I)f\| + b\|R_\lambda(A)f\|$$
$$\leq (2a + b/\lambda)\|f\|$$

and the result follows when we take $\lambda > b/(1 - 2a)$.

Case II ($1/2 \leq a < 1$).

First, let $0 \leq \alpha < 1$. Then for all $f \in D_A$, using (6.1.1), we have

$$\|(A + \alpha B)f\| \geq \|Af\| - \alpha\|Bf\|$$
$$\geq \|Af\| - \|Bf\|$$
$$\geq (1 - a)\|Af\| - b\|f\|.$$

Now choose $n \in \mathbb{N}$ so that $a/n < (1 - a)/4$ (e.g., if $a = 1/2$, then any $n \geq 5$ will do). Now by (6.1.1) again, we have

$$\left\| \frac{1}{n} B f \right\| \leq \frac{a}{n} \|Af\| + \frac{b}{n} \|f\|$$

$$\leq \frac{1-a}{4} \|Af\| + \frac{b}{n} \|f\|$$

$$\leq \frac{1}{4} \|(A + \alpha B)f\| + b \left(\frac{1}{n} + \frac{1}{4} \right).$$

So the operator B/n is relatively bounded with respect to $A + \alpha B$ and we can use the result of case I to conclude that if $A + \alpha B$ generates a contraction semigroup, then so does $A + (\alpha + 1/n)B$. Now put $\alpha = 0$. Then we see immediately that $A + 1/nB$ generates a contraction semigroup. Next, put $\alpha = 1/n$, and we find that $A + 2/nB$ also generates a contraction semigroup. Now repeat the argument n times to conclude, as required, that $A + B = A + (n/n)B$ generates a contraction semigroup.   □

For interesting examples of relatively bounded operators, where both A and B are unbounded, see Kato [59], pp. 191–4. We now want to consider the case where B is a bounded operator. It is clear from Theorem 6.1.1, and remarks after (6.1.1), that if A generates a contraction semigroup, and B is bounded and dissipative, then $A + B$ generates a contraction semigroup. In order to drop the dissipativity assumption on B and to move from contraction semigroups to the more general C_0 case, we first require a technical lemma.

Lemma 6.1.2 *Let* $(T_t, t \geq 0)$ *be a* C_0-*semigroup on* E *having generator* A, *so that there exists* $M \geq 1, a \in \mathbb{R}$ *with* $\|T_t\| \leq Me^{at}$ *for all* $t \geq 0$. *Then there exists an equivalent norm* $\| \cdot \|_1$ *on* E *so that* $(S_t, t \geq 0)$ *is a contraction semigroup on* $(E, \| \cdot \|_1)$ *where* $S_t := e^{-at}T_t$ *for all* $t \geq 0$. *Furthermore, the generator of* $(S_t, t \geq 0)$ *is* $A - aI$ *with domain* D_A.

Proof. It is easy to check that $(S_t, t \geq 0)$ is a C_0-semigroup with generator $A - aI$ and $\|S_t\| \leq M$ for all $t \geq 0$. Define $\|f\|_1 := \sup_{t \geq 0} \|S_t f\|$ for all $f \in D_A$. Then we have

$$\|f\| \leq \|f\|_1 \leq M\|f\|,$$

where the left-hand inequality comes from the fact that $\|f\| = \|S_0 f\|$. Then S_t is a contraction for all $t \geq 0$ as for all $f \in E$,

$$\|S_t f\|_1 = \sup_{r \geq 0} \|S_{t+r} f\| \leq \sup_{r \geq 0} \|S_r f\| = \|f\|_1.$$ □

Theorem 6.1.3 *If* A *generates a* C_0-*semigroup on* E *and* $B \in \mathcal{L}(E)$, *then* $A + B$ *generates a* C_0-*semigroup on* E.

Proof. Since for all $a \in \mathbb{R}$, we have $A + B = (A - aI) + (B + aI)$, we may effectively replace A by $A - aI$ and assume that $||e^{tA}|| \leq M$ for all $t \geq 0$, where $M \geq 1$. If we now use the norm $|| \cdot ||_1$ from Lemma 6.1.2, we see that A generates a contraction semigroup on $(E, || \cdot ||_1)$. Next we observe that $B - ||B||_1 I$ is dissipative, since for all $(f, \phi) \in E \times E'$ with $\langle f, \phi \rangle_1 = ||f||_1 = ||\phi||_1 = 1$, we have

$$\Re \langle (B - ||B||_1) f, \phi \rangle_1 = \Re \langle Bf, \phi \rangle_1 - ||B||_1 \langle f, \phi \rangle_1$$
$$\leq ||B||_1 (1 - \langle f, \phi \rangle_1) = 0.$$

Since for all $f \in D_A$,

$$||(B - ||B||_1) f||_1 \leq 2||B||_1 ||f||_1,$$

we may apply Theorem 6.1.1 with $a = 0$ and $b = 2||B||_1$ to deduce that $A + B - ||B||_1 I$ generates a contraction semigroup on $(E, || \cdot ||_1)$. Then it is easily verified that $A + B = (A + B - ||B||_1 I) + ||B||_1$ generates a C_0-semigroup on $(E, || \cdot ||)$. $\qquad \square$

It is worth pointing out that Theorem 6.1.3 is unnecessary if A and B commute. Then we can just write $e^{t(A+B)} = e^{tA} e^{tB}$ for all $t \geq 0$. More generally, although the proof of Theorem 6.1.3 is quite elegant, it could be argued that the route to the conclusion (via Theorem 6.1.1) is somewhat indirect. There is a more direct way of obtaining the result, which can be found in, e.g., Davies [27], pp. 339–41. It involves defining the required semigroup as an infinite series

$$e^{t(A+B)} f := e^{tA} f + \int_0^t e^{(t-s)A} B e^{sA} f \, ds$$
$$+ \int_0^t \int_0^s e^{(t-s)A} B e^{(s-r)A} B e^{rA} f \, ds dr$$
$$+ \int_0^t \int_0^s \int_0^r e^{(t-s)A} B e^{(s-r)A} B e^{(r-u)A} B e^{uA} f \, ds dr du + \cdots,$$

for all $t \geq 0$, $f \in E$. Checking the semigroup property (S1) then becomes a tedious exercise in series manipulations.

6.1.2 Analytic Semigroups

Perturbation theory is important for applications, and it is helpful to identify classes of semigroups for which there are powerful results in this regard. Contraction semigroups are one such class, as we have already seen in this chapter. Self-adjoint semigroups are another, and we have already pointed out in Chapter 4 the significance for quantum mechanics of perturbing the Laplacian by a potential (see also Problem 6.4 for a guided proof of the Kato–Rellich theorem). Another important class is that of *analytic semigroups*, which we now introduce. Vaguely speaking, these are C_0-semigroups that may be analytically continued to a sector in the complex plane in such a way as to preserve the semigroup property, and to be able to make use of the deep and beautiful insights of complex function theory. To be precise, for $0 < \theta \le \pi/2$, define the sector

$$\Sigma_\theta := \{z \in \mathbb{C}; z \ne 0; |\arg(z)| < \theta\}.$$

We say that a family of operators $(T_z; z \in \Sigma_\theta \cup \{0\})$ is an *analytic semigroup of angle* θ if

1. $T_0 = I$ and $T_z T_w = T_{z+w}$, for all $z, w \in \Sigma_\theta$.
2. The mapping $z \to \langle T_z f, \phi \rangle$ is analytic on Σ_θ, for all $f \in E, \phi \in E'$.
3. $\lim_{z \in \Sigma_{\theta'}; z \to 0} T_z f = f$, for all $f \in E$ and all $0 < \theta' < \theta$.

It follows that $(T_t, t \ge 0)$ is a C_0-semigroup and so has a generator A in the usual sense. It is common to call A the generator of the analytic semigroup $(T_z; z \in \Sigma_\theta \cup \{0\})$.

We say that an analytic semigroup is *bounded* if the mapping $z \to ||T(z)||$ is bounded on $\Sigma_{\theta'}$ for all $0 < \theta' < \theta$. The class of analytic semigroups is large enough to include all self-adjoint contraction semigroups in Hilbert space.

We will not develop the theory of analytic semigroups herein. Accounts can be found in the standard references, such as Engel and Nagel [32], Chapter 2.4, pp. 90–104 or Davies [25], section 2.6, pp. 59–67. We will be content with stating the following result, whose proof may be found in Engel and Nagel [32], Theorem 2.10, pp. 130–2.

Theorem 6.1.4 *Let A be the generator of an analytic semigroup $(T_z; z \in \Sigma_\theta \cup \{0\})$. Then there exists $\rho > 0$ such that $A + B$ generates an analytic semigroup for every relatively bounded B which has A-bound $a_0 < \rho$.*

6.2 The Lie–Kato–Trotter Product Formula

For $d \times d$-matrices A and B, there is a very beautiful formula due to Sophus Lie:

$$e^{A+B} = \lim_{n \to \infty} (e^{A/n} e^{B/n})^n, \tag{6.2.2}$$

for all $t \in \mathbb{R}$. We seek to extend this to contraction semigroups, and we will follow Davies [25], pp. 90–3. We first establish two lemmas:

Lemma 6.2.5 *If C is a contraction in E, then $(e^{t(C-1)}, t \geq 0)$ is a norm-continuous contraction semigroup for which*

$$||e^{n(C-1)}f - C^n f|| \leq \sqrt{n}||Cf - f||,$$

for all $f \in E, n \in \mathbb{N}$.

Proof. Since $C - 1$ is bounded, it generates a norm-continuous semigroup by Theorem 1.4.21. This is a contraction semigroup, as for all $t \geq 0$,

$$||e^{t(C-1)}|| \leq e^{-t} \left|\left| \sum_{n=0}^{\infty} \frac{t^n}{n!} C^n \right|\right| \leq e^{-t} \sum_{n=0}^{\infty} \frac{t^n}{n!} = 1.$$

To obtain the required estimate, observe that for each $n \in \mathbb{N}$, $f \in E$:

$$||e^{n(C-1)}f - C^n f||$$

$$\leq e^{-n} \left|\left| \sum_{m=0}^{\infty} \frac{n^m}{m!} (C^m f - C^n f) \right|\right|$$

$$\leq e^{-n} \left|\left| \sum_{m=0}^{\infty} \frac{n^m}{m!} (C^{|m-n|} f - f) \right|\right|$$

$$= e^{-n} \left|\left| \sum_{m=0}^{\infty} \frac{n^m}{m!} (C^{|m-n|} f - C^{|m-n|-1} f + C^{|m-n|-1} f \right.\right.$$

$$\left.\left. - C^{|m-n|-2} f + \cdots - f) \right|\right|$$

$$\leq e^{-n} ||Cf - f|| \sum_{m=0}^{\infty} \frac{n^m}{m!} |m - n|$$

$$= ||Cf - f|| \sum_{m=0}^{\infty} e^{-n/2} \frac{n^{m/2}}{\sqrt{m!}} . e^{-n/2} \frac{n^{m/2}}{\sqrt{m!}} |m - n|$$

$$\leq ||Cf - f|| \left(\sum_{m=0}^{\infty} e^{-n} \frac{n^m}{m!} (m - n)^2 \right)^{1/2},$$

where we have used Cauchy's inequality to obtain the last line. But

$$\sum_{m=0}^{\infty} \frac{n^m}{m!} (m - n)^2 = \sum_{m=0}^{\infty} \frac{n^m}{m!} [m(m - 1) + m - 2mn + n^2] = ne^n,$$

and the result follows. $\qquad\square$

Lemma 6.2.6 (Chernoff's Product Formula) *Let A be the infinitesimal generator of a contraction semigroup $(T_t, t \geq 0)$ in E with a given core D and let F be a mapping from $[0, \infty)$ to $\mathcal{L}(E)$ such that $F(t)$ is a contraction for all $t \geq 0$, with $F(0) = I$. If we have that for all $f \in D$,*

$$\lim_{t \to 0} \frac{F(t)f - f}{t} = Af,$$

then $T_t f = \lim_{n \to \infty} F(t/n)^n f$ for all $f \in E$.

Proof. For each $t \geq 0$, define $A_n(t) := \frac{n}{t}(F(t/n) - 1)$. Then $A_n(t)$ is a bounded linear operator in E for which $\lim_{n \to \infty} A_n(t)f = Af$ for all $f \in D$. Then by a similar argument to that used in the proof of Theorem 2.2.4 we have $\lim_{n \to \infty} e^{tA_n(t)} f = T_t f$, and the convergence is uniform on bounded intervals.

By Lemma 6.2.5, we have for all $t \geq 0$,

$$
\begin{aligned}
||e^{tA_n(t)} f - F(t/n)^n f|| &= || \exp\{n(F(t/n) - 1)\}f - F(t/n)^n f|| \\
&= \sqrt{n}||F(t/n)f - f|| \\
&= t/\sqrt{n}||A_n(t)f|| \to 0 \text{ as } n \to \infty.
\end{aligned}
$$

Hence we have $\lim_{n \to \infty} ||T_t f - F(t/n)^n f|| = 0$, and this extends to all $f \in E$ by density of D (see Problem 6.2). $\qquad\square$

We can now establish the main result of this section:

Theorem 6.2.7 (Lie–Kato–Trotter) *If A, B and C are each generators of contraction semigroups in E, D is a core for C and we have*

$$Cf = Af + Bf,$$

for all $f \in D$, then for all $t \geq 0$,

$$e^{tC} f = \lim_{n \to \infty} (e^{tA/n} e^{tB/n})^n f.$$

Proof. Define the contractions $F(t) = e^{tA} e^{tB}$ for all $t \geq 0$. Then $F(0) = I$ and if $f \in D$ then

$$
\begin{aligned}
\lim_{t \to 0} \frac{1}{t}(F(t)f - f) &= \lim_{t \to 0} \left\{ e^{tA} \frac{1}{t}(e^{tB}f - f) \right\} + \frac{1}{t}(e^{tA}f - f) \\
&= Bf + Af = Cf,
\end{aligned}
$$

and the result now follows from Lemma 6.2.6. $\qquad\square$

6.3 The Feynman–Kac Formula

We return to the Schrödinger equation (4.3.4). Richard Feynman developed an intriguing method of solving this equation, which is nowadays called a *Feynman integral* in his honour. This is used extensively in modern physics, but it has perplexed mathematicians as it cannot be made into a fully rigorous object, although there have been several interesting attempts. For more on this, see the discussion on pp. 54–5 in Goldstein [39], and also Albeverio and Høegh-Krohn [4], The problem is that the integral is defined in terms of a formal measure on the space of all paths that a quantum particle might explore. Mark Kac [57] noticed that when you replaced time t with imaginary time $-it$, then the formal measure on path space of Feynman is effectively transformed into Wiener measure, and Feynman's procedure leads to a perturbation of the usual diffusion of a Brownian particle by a potential term. To be precise, we seek to solve the following parabolic equation in $\mathbb{R}^d$:

$$\frac{\partial \psi}{\partial t} = \Delta \psi - V \psi; \, \psi(0) = f, \tag{6.3.3}$$

where $f \in C_0^2(\mathbb{R}^d)$, and for simplicity we will make the assumption that $V \in C(\mathbb{R}^d)$. We also require $V \geq 0$. This ensures that all terms we consider which involve V are integrable, but is also desirable on physical grounds, if we think of V as a potential term.[1]

If we put $V = 0$, then the solution to (6.3.3) is given by

$$\phi(t, x) = \mathbb{E}(f(x + B(t))),$$

as was shown in section 3.1.2. Kac argued that the solution to (6.3.3) with non-zero V should be

$$\psi(t, x) = \mathbb{E}\left(\exp\left\{-\int_0^t V(x + B(s))ds\right\} f(x + B(t))\right). \tag{6.3.4}$$

We'll give two proofs, one based on the Lie–Kato–Trotter product formula and properties of Wiener measure, and the other using Itô's formula, for those readers who know some stochastic analysis.

6.3.1 The Feynman–Kac Formula via the Lie–Kato–Trotter Product Formula

In (6.2.2), we take $E = L^2(\mathbb{R}^d)$, $A = \Delta$ and $Bf(x) = -V(x)f(x)$ for $f \in L^2(\mathbb{R}^d)$ and $x \in \mathbb{R}^d$. Then both of these operators generate C_0-semigroups with, for each $t > 0$,

---

[1] The Laplacian Δ is (essentially) minus the kinetic energy.

$$e^{tA} f(x) = \int_{\mathbb{R}^d} f(y) \gamma_t(x - y) dy \text{ and } e^{tB} f(x) = e^{-tV(x)} f(x),$$

for all $f \in L^2(\mathbb{R}^d)$, $x \in \mathbb{R}^d$, where γ_t is the heat kernel. By using induction we have that for all $n \in \mathbb{N}$, $x \in \mathbb{R}^d$,

$$(e^{tA/n} e^{tB/n})^n f(x)$$
$$= \int_{\mathbb{R}^d} \int_{\mathbb{R}^d} \cdots \int_{\mathbb{R}^d} e^{-t/n \sum_{j=1}^n V(y_j)} \gamma_{t/n}(y_1 - x) \gamma_{t/n}(y_2 - y_1) \cdots$$
$$\gamma_{t/n}(y_n - y_{n-1}) f(y_n) dy_1 dy_2 \ldots dy_n. \tag{6.3.5}$$

To proceed further we must introduce a Gaussian measure on an infinite-dimensional space, the celebrated *Wiener measure* on the path space:

$$\Omega = \{\omega : [0, \infty) \to \mathbb{R}^d, \omega \text{ continuous with } \omega(0) = 0\}.$$

We equip Ω with the σ-algebra $\mathcal{F}$ which is defined to be smallest σ-algebra containing all *cylinder sets* of the form

$$I_A^{t_1, t_2, t_n} = \{\omega \in \Omega; (\omega(t_1), \ldots, \omega(t_n)) \in A\},$$

where $A \in \mathcal{B}(\mathbb{R}^{dn})$, $0 \le t_1 < \cdots < t_n < \infty$, and $n \in \mathbb{N}$. The action of Wiener measure P on cylinder sets is given by

$$P(I_A^{t_1, t_2, t_n}) = \int_A \gamma_{t_1}(y_1) \gamma_{t_2 - t_1}(y_2 - y_1) \cdots \gamma_{t_n - t_{n-1}}(y_n - y_{n-1}) dy_1 dy_2 \ldots dy_n.$$

We should prove that P really is a measure on $(\Omega, \mathcal{F})$, but this is covered in many good texts (e.g., Billingsley [12]), so let us refer the reader to these. We note two key facts about Wiener measure:

Fact 1. Once Wiener measure is obtained, we get an existence proof for Brownian motion. More precisely, Brownian motion is defined on the probability space $(\Omega, \mathcal{F}, P)$ by the prescription

$$B(t)\omega = \omega(t),$$

for all $t \ge 0$, $\omega \in \Omega$.

Fact 2. Given any measurable function $g : \mathbb{R}^{dn} \to \mathbb{R}$ and $0 \le t_1 < \cdots < t_n < \infty$, where $n \in \mathbb{N}$, we obtain a measurable function $F_g^{t_1, t_2, t_n} : \Omega \to \mathbb{R}$ by the prescription, for each $\omega \in \Omega$,

$$F_g^{t_1, t_2, t_n}(\omega) = g(\omega(t_1), \ldots, \omega(t_n)).$$

$F_g^{t_1, t_2, t_n}$ is called a *cylinder function*, and it is clear that a simple cylinder function is just a finite linear combination of indicator functions of cylinder sets.

Now let $a \in \mathbb{R}^d$, and for each $t \geq 0$, define $B_a(t) := a + B(t)$ to be Brownian motion with drift. We introduce the translation map

$$(\tau_a g)(y_1, \ldots y_n) = g(y_1 + a_1, \ldots, y_n + a_n)$$

for each $y_1, \ldots, y_n \in \mathbb{R}^d$. We then have that

$$\int_\Omega F_{\tau_a g}^{t_1, t_2, t_n}(\omega) P(d\omega) = \mathbb{E}(g(B_a(t_1), \ldots, B_a(t_n)))$$

$$= \int_{\mathbb{R}^d} \int_{\mathbb{R}^d} \cdots \int_{\mathbb{R}^d} g_{y_1, \ldots, y_n} \gamma_{t_1}(y_1 - a) \gamma_{t_2 - t_1}(y_2 - y_1) \ldots$$
$$\gamma_{t_n - t_{n-1}}(y_n - y_{n-1}) dy_1 dy_2 \ldots dy_n. \qquad (6.3.6)$$

If we compare (6.3.5) with (6.3.6), we see that the former may now be interpreted as the integral of a cylinder function where $a = x, t_j = jt/n$ for $j = 1, \ldots, n$, and

$$g(y_1, \ldots y_n) = e^{-t/n \sum_{j=1}^n V(y_j)} f(y_n).$$

Let us write the corresponding cylinder function $F_g^{t_1, t_2, t_n}$ as $H_{V, f}^{(n)}$. We are now ready to apply the Lie–Kato–Trotter product formula. By dominated convergence we have

$$\lim_{n \to \infty} (e^{tA/n} e^{tB/n})^n f(x) = \int_\Omega \lim_{n \to \infty} H_{V, \tau_x f}^{(n)}(\omega) P(d\omega).$$

However, as for all $\omega \in \Omega, t \geq 0, \lim_{n \to \infty} \sum_{j=1}^n V(\omega(jt/n))t/n = \int_0^t V(\omega(s))ds$, we conclude that

$$\lim_{n \to \infty} H_{V, \tau_x f}^{(n)}(\omega) = \exp\left\{-\int_0^t V(x + \omega(s))ds\right\} f(x + \omega(t))$$

$$= \exp\left\{-\int_0^t V(x + B(s))ds\right\} f(x + B(t))(\omega),$$

by Fact 1, and (6.3.4) then follows from (6.3.6).

6.3.2 The Feynman–Kac Formula via Itô's Formula

In the case where $V \in C_b(\mathbb{R}^d)$, we also present a swift probabilistic argument that requires only Itô's formula (that we met in section 3.1.2) and the notion of a martingale. We will also take $d = 1$ to simplify matters further. Fix $t > 0$ and $x \in \mathbb{R}$ and define a stochastic process $(Y(s), 0 \leq s \leq t)$ by the prescription

$$Y(s) = \exp\left\{-\int_0^s V(x + B(u))du\right\}\psi(t - s, x + B(s)).$$

Now writing $R(s) = \int_0^t V(x + B(s))ds$, we find by using Itô's formula that

$$\begin{aligned}
dY(s) &= -e^{-R(s)}V(x + B(s))\psi(t - s, x + B(s))ds \\
&\quad - e^{-R(s)}\frac{\partial\psi(t - s, B(s) + x)}{\partial s}ds \\
&\quad + e^{-R(s)}\frac{\partial\psi(t - s, B(s) + x)}{\partial x}dB(s) \\
&\quad + e^{-R(s)}\frac{\partial^2\psi(t - s, B(s) + x)}{\partial x^2}ds \\
&= e^{-R(s)}\frac{\partial\psi(t - s, B(s) + x)}{\partial x}dB(s),
\end{aligned}$$

since ψ satisfies the PDE (6.3.3). So we can represent the process Y as a stochastic integral:

$$Y(s) = Y(0) + \int_0^s e^{-R(u)}\frac{\partial\psi(t - u, B(u) + x)}{\partial x}dB(u),$$

which is a martingale. But then we have

$$\begin{aligned}
Y(0) &= \psi(t, x) \\
&= \mathbb{E}(Y(t)) \\
&= \mathbb{E}(e^{-R(t)}\psi(0, x + B(t))) \\
&= \mathbb{E}(e^{-R(t)}f(x + B(t))),
\end{aligned}$$

and we have derived (6.3.4).

The Feynman–Kac formula may be significantly generalised, in that Brownian motion may be replaced by a general Feller process X (see Chapter 7 below), and the potential V need not be non-negative, but must satisfy an *abstract Kato condition* of the form

$$\lim_{t \to 0}\sup_{x \in \mathbb{R}^d}\int_0^t T_s(|V|(x))ds = 0,$$

where $(T_t, t \geq 0)$ is the Feller semigroup associated to X. For further discussion, see section 4.4 of Böttcher et al. [17], pp. 108–10, and references therein.

6.4 Exercises for Chapter 6

1. Let $(T_t, t \geq 0)$ be a contraction semigroup having resolvent R_λ for $\lambda > 0$. Use Chernoff's product formula to show that for all $t \geq 0$, $f \in E$,

$$T_t f = \lim_{n \to \infty} \left(\frac{n}{t} R_{n/t} \right)^n f.$$

This is called the *Post–Widder inversion formula*. Hint: Define

$$F(t) = \begin{cases} I & \text{if } t = 0, \\ \frac{1}{t} R_{1/t} & \text{if } t > 0. \end{cases}$$

(For a generalisation to C_0-semigroups, see Engel and Nagel [32], pp. 151–3.)

2. Complete the density argument at the end of the proof of Chernoff's product formula (Lemma 6.2.6).

3. *Based on Lemma 2.4 in Engel and Nagel [32], p. 126.*
   If A is a closed linear operator acting in a Banach space E, and B is relatively bounded with respect to A with $a < 1$, show that $A + B$ is closed with domain D_A.
   [Hint: Show that the graph norm of $A + B$ is equivalent to that of A.]

4. Here we will establish the main part of the *Kato–Rellich theorem*, which states that if A is self-adjoint in a Hilbert space H, B is symmetric and also relatively bounded with respect to A with $a < 1$, then $A + B$ is self-adjoint on D_A. We use a (easily established) variant of the basic criterion of self-adjointness (Problem 4.2), and seek to show that $\text{Ran}(A + B \pm i\lambda I) = H$, for some $\lambda > 0$, and the result follows. In fact we just aim to establish $\text{Ran}(A + B + i\lambda I) = H$ below as the other result is proved in the same way.

   (a) For $\phi \in D_A$, $\lambda > 0$, compute $||(A + i\lambda I)\phi||^2$ and then let $\phi = (A + i\lambda I)^{-1}\psi$. From here show that

   $$||A(A + i\lambda I)^{-1}|| \leq 1 \text{ and } ||(A + i\lambda I)^{-1}|| \leq 1/\lambda.$$

   (b) Define $X = B(A + i\lambda I)^{-1}$ and show that

   $$||X\psi|| \leq (a + b/\lambda)||\psi||.$$

   (c) Show that if λ is chosen to be sufficiently large in (b), then $||X|| < 1$ and so[2] $-1 \notin \sigma(X)$.

[2] Use the fact that the *spectral radius* of $C \in \mathcal{L}(H)$, which is defined to be $\sup_{\mu \in \sigma(C)} |\mu|$, is also equal to $\lim_{n \to \infty} ||C^n||^{\frac{1}{n}}$; see, e.g., Reed and Simon [76], Theorem VI.6, p. 192.

(d) Deduce that $\text{Ran}(I + X) = H$.

(e) Use the fact that

$$(I + X)(A + i\lambda I)\phi = (A + B + i\lambda I)\phi,$$

for all $\phi \in D_A$ all to conclude the proof.

This proof is based on that given in Reed and Simon [77], Theorem X.12, pp. 162–3.

7

Markov and Feller Semigroups

7.1 Definitions of Markov and Feller Semigroups

We return to some of the themes of Chapter 3, but in a more general context. Assume we are given a measurable space (S, Σ), and a family $(p_t, t \geq 0)$ where for each $t \geq 0$, $p_t : S \times \Sigma \to [0, 1]$ satisfies the following conditions for each $t \geq 0$:

1. The mapping $p_t(\cdot, A)$ is a $(\Sigma, \mathcal{B}([0, 1]))$ measurable function, for each $A \in \Sigma$.
2. The mapping $A \to p_t(x, A)$ is a probability measure for all $x \in S$.

We then say that $(p_t, t \geq 0)$ is a *probability kernel*. For example, take (S, Σ) to be $(\mathbb{R}^d, \mathcal{B}(\mathbb{R}^d))$ and $(\mu_t, t \geq 0)$ to be a convolution semigroup. Then $p_t(x, A) = \mu_t(A - x)$ defines a probability kernel, where $A - x := \{y - x; y \in A\}$.

We have seen that convolution semigroups in $\mathbb{R}^d$ give rise to C_0-semigroups on the space $C_0(\mathbb{R}^d)$. We have imposed no topology on S, so we cannot define any analogue of that space. Of course we could equip S with a topology, but we prefer to remain at our current level of generality. Let us instead work in the Banach space $B_b(S)$ of bounded measurable functions from S to $\mathbb{R}$, and define a family of linear operators $(T_t, t \geq 0)$ by

$$(T_t f)(x) = \int_S f(y) p_t(x, dy), \qquad (7.1.1)$$

for each $t \geq 0$, $f \in B_b(S)$, $x \in S$. It is easy to see that T_t is a contraction.

We say that the kernel (p_t) is a *Markov kernel* if $(T_t, t \geq 0)$ is an AO (contraction) semigroup in that (S1) and (S2) both hold, and we call

129

$(T_t, t \geq 0)$ a *Markov semigroup*. The requirement that $T_0 = I$ immediately yields $p_t(x, \cdot) = \delta_x$. Now if (S1) holds, we have

$$
\int_S f(z) p_{s+t}(x, dz) = (T_{s+t} f)(x)
$$
$$
= T_s(T_t f)(x)
$$
$$
= \int_S (T_t f)(y) p_s(x, dy)
$$
$$
= \int_S \int_S f(z) p_t(y, dz) p_s(x, dy).
$$

Now take $f = \mathbf{1}_A$ where $A \in \Sigma$, to find that:

$$
p_{s+t}(x, A) = \int_S p_t(y, A) p_s(x, dy), \qquad (7.1.2)
$$

and these are the famous *Chapman–Kolmogorov equations* which we have just shown to be necessary conditions for a probability kernel to be Markov. It is easy to check that they are also sufficient.

When we studied convolution semigroups, we showed that there was a class of stochastic processes, called Lévy processes, which are associated to them. The same is true of Markov kernels. Let $(\Omega, \mathcal{F}, P)$ be a probability space and $(\mathcal{F}_t, t \geq 0)$ be a *filtration* of $\mathcal{F}$ so that for each $t \geq 0$, $\mathcal{F}_t$ is a sub-σ-algebra of $\mathcal{F}$ with $\mathcal{F}_s \subseteq \mathcal{F}_t$ whenever $0 \leq s \leq t$. A stochastic process $X = (X_t, t \geq 0)$ taking values in S is said to be *adapted* to the given filtration if X_t is $\mathcal{F}_t$–measurable for all $t \geq 0$. Finally, X is said to be a *(time-homogeneous) Markov process* if for all $f \in B_b(\mathbb{R}^d)$, $s, t \geq 0$ it satisfies the *Markov property*

$$
\mathbb{E}(f(X_{s+t})|\mathcal{F}_s) = (T_t f)(X_s), \qquad (7.1.3)
$$

where for all $x \in \mathbb{R}^d$, $T_t f(x) := \mathbb{E}(f(X_t)|X_0 = x)$. Then it can be shown that the associated Markov kernel is given by

$$
p_t(x, A) = P(X_t \in A | X_0 = x) = (T_t \mathbf{1}_A)(x).
$$

This is interpreted as the *transition probability* for the process to take values in A at time t, given that it starts out at the point x at time 0. The Markov semigroup $(T_t, t \geq 0)$, as given by (7.1.1), is sometimes called the *transition semigroup* for the process.

The range of examples of Markov processes is very broad, and we will not seek to explore these here. Certainly, the solution of a stochastic differential equation with suitably regular coefficients that is driven by Brownian motion, or more generally a Lévy process, will be a Markov process. For details, see Chapter 3 of Applebaum [6].

We make a few remarks about the meaning of $\mathbb{E}(\cdot|X_0 = x)$. This is not to be interpreted as a conditional expectation in the usual sense; indeed, we may well have $P(X_0 = x) = 0$. Let us suppose that we have some choice over the initial law μ of the random variable X_0, and let $(X_t^x, t \geq 0)$ be the Markov process obtained by taking $\mu = \delta_x$. Then we write $\mathbb{E}(f(X_t)|X_0 = x)$ as a suggestive shorthand notation for $\mathbb{E}(f(X_t^x))$. Why do we do this? One reason (not a good one) is that it is well-established notation within the community. Another is that there is some economy of thought in thinking of a single Markov process that we "condition" to start at $X_0 = x$ (a.s.), rather than a field of Markov processes $(X_t^x, t \geq 0, x \in S)$. Note that some authors shift the x-dependence onto the measure and replace P with a family of measures $(P^x, x \in S)$ defined on $(\Omega, \mathcal{F})$. They would then write $\mathbb{E}^x(f(X_t))$ instead of $\mathbb{E}(f(X_t)|X_0 = x)$ (see, e.g., Chapter 3 of Rogers and Williams [79]).

So far we have said nothing about (S3), so let us remedy that now. We assume that S is equipped with a topology[1] so that it is a locally compact Hausdorff space and that Σ is its Borel σ–algebra $\mathcal{B}(S)$. A Markov process taking values in S is said to be a *Feller process* if the associated AO (contraction) semigroup $(T_t, t \geq 0)$ is in fact a C_0-semigroup on $C_0(S)$. This requires that we have

(F1) $T_t : C_0(S) \subseteq C_0(S)$ for all $t \geq 0$.
(F2) $\lim_{t\to 0} \|T_t f - f\|_\infty = 0$ for all $f \in C_0(S)$.

If X is a Feller process on S, then its transition semigroup is called a *Feller semigroup*.[2] Again the class of Feller processes is very wide. It contains all Lévy processes and a large class of processes that are obtained by solving stochastic differential equations. For a simple example of a Feller process that is not a Lévy process, see the construction of *Feller's pseudo-Poisson process* in Applebaum [6], pp. 185–6.

The definition of a Feller semigroup that we have given is inherently probabilistic. Can we find an alternative analytic characterisation?

First, let us give some definitions. A C_0-semigroup $(T_t, t \geq 0)$ defined on $C_0(S)$ is said to be *positive* if $f \in C_0(S)$ with $f \geq 0$ implies that $T_t f \geq 0$ for all $t \geq 0$. It is said to be *conservative* if for each $t \geq 0$, the operator T_t has an extension to $B_b(S)$ so that $T_t 1 = 1$.

---

[1] It is fine to just take $S \subseteq \mathbb{R}^d$ if you have not studied topology yet.
[2] It is worth bearing in mind that there are a multitude of different uses of the term "Feller semigroup" in the literature. For example, Jacob [53] calls any positive contraction semigroup on $C_0(S)$ a Feller semigroup (Definition 4.1.4, p. 250). Rogers and Williams write, "Every author has his or her own definition of a 'Feller semigroup'". [79], p. 255.

Theorem 7.1.1 *A contraction semigroup* $(T_t, t \geq 0)$ *on* $C_0(S)$ *is a Feller semigroup if and only if it is both positive and conservative.*

Proof. Necessity is obvious. For sufficiency, suppose that $(T_t, t \geq 0)$ is a conservative positive contraction semigroup. For each $t \geq 0$, $x \in S$, define linear functionals $L_{t,x}$ on $C_0(S)$ by $L_{t,x}f := (T_t f)(x)$. Then $L_{t,x}$ is positivity preserving, and so by the Riesz lemma (Theorem G.0.1), it determines a measure, which we write as $p_t(x, \cdot)$ on $(S, \mathcal{B}(S))$. Thus we have

$$T_t f(x) = \int_S f(y) p_t(x, dy),$$

for all $f \in C_0(S)$. It is easy to see that $|L(t, x)(f)| \leq \|f\|_\infty$, for all $f \in C_0(S)$, and so we have a bounded linear functional. Hence the measures $p_t(x, \cdot)$ are finite (see, e.g., Theorem 7.3.5 on pp. 220–1 of Cohn [22]). The integral representation that we have just defined clearly gives an extension of T_t to $B_b(S)$ and we thus have

$$p_t(x, S) = T_t \mathbf{1}_S(x) = T_1(1)(x) = 1,$$

so that $p_t(x, \cdot)$ is indeed a probability measure. Measurability of $x \to p_t(x, A) = T_t \mathbf{1}_A(x)$ is then immediate from the definition of conservativity that we have given. The fact that $(p_t(x, \cdot))$ is a Markov kernel then follows from the semigroup property as in the discussion at the beginning of this chapter. We then obtain a consistent family of probability measures (see Appendix 3) by defining, for $m \in \mathbb{N}$, $A \in B_b(S)^m$,

$$
\begin{aligned}
p_{t_1, t_2, \ldots, t_m}(A) \\
= \int_A p_{t_m - t_{m-1}}(y_{m-1}, dy_m) p_{t_{m-1} - t_{m-2}}(y_{m-2}, dy_{m-1}) \cdots \\
p_{t_2 - t_1}(y_1, dy_2) p_{t_1}(x, dy_1),
\end{aligned}
$$

where $0 < t_1 < \cdots t_m < \infty$. The construction of a Markov process, starting at x (almost surely) and having these as transition probabilities, is then a standard argument, using Kolmogorov's existence theorem as described in Appendix 3 (see also Theorem 3.1.7 in Applebaum [6]).   □

Note. Theorem 7.1.1 has rather an easy proof, but in practice, the burden has been placed on showing that the semigroup operators really do extend to $B_b(S)$. If you do not make this assumption, you can still prove that the mappings $x \to p_t(x, A)$ are measurable by using monotone class arguments – see the discussion on p. 7 of Bottcher et al. [17].

Our next goal is to seek to gain some insight into the structure of the generator of a Feller semigroup.

7.2 The Positive Maximum Principle

7.2.1 The Positive Maximum Principle and the Hille–Yosida–Ray Theorem

From now on let S be a locally compact, Hausdorff separable space and A be a linear operator acting on $C_0(S)$ with domain D_A. The operator A is said to satisfy the *positive maximum principle* (PMP) if, given any $f \in D_A$ which is such that there exists $x_0 \in S$ for which $f(x_0) = \sup_{x \in S} f(x) \geq 0$, then $Af(x_0) \leq 0$. For example, take $S = \mathbb{R}$ and $Af = f''$ on $D_A = C_c^2(\mathbb{R})$. The importance for us of the PMP comes from the following three results:

Proposition 7.2.2 *If A is the generator of a Feller semigroup in $C_0(S)$, then A satisfies the PMP.*

Proof. By Theorem 7.1.1, there exists a transition kernel p_t so that for all $f \in C_0(S), t \geq 0, x \in S$,

$$T_t f(x) = \int_S f(y) p_t(x, dy).$$

Now let $f \in D_A$ and suppose there exists $x_0 \in S$ for which $f(x_0) = \sup_{x \in S} f(x) \geq 0$. Then

$$Af(x_0) = \lim_{t \to 0} \frac{1}{t} \int_S (f(y) - f(x_0)) p_t(x_0, dy) \leq 0,$$

and the result is established. $\qquad\square$

Next we look at analytic implications of the PMP.

Proposition 7.2.3 *If A satisfies the PMP on $C_0(S)$, then for all $\lambda > 0$, $f \in D_A$*

$$||(\lambda I - A)f||_\infty \geq \lambda ||f||_\infty. \qquad (7.2.4)$$

Proof. Clearly $|f| \in C_0(S)$ and so $|f|$ attains its supremum at some $x_0 \in S$, so we have $|f(x_0)| = ||f||_\infty$. Assume that $f(x_0) \geq 0$ (if not, we may replace f with $-f$). Then $Af(x_0) \leq 0$ and, for all $\lambda > 0$,

$$
\begin{aligned}
||(\lambda I - A)f||_\infty &= \sup_{x \in S} |\lambda f(x) - Af(x)| \\
&\geq \lambda f(x_0) - Af(x_0) \\
&\geq \lambda f(x_0) = \lambda ||f||_\infty. \qquad\square
\end{aligned}
$$

Remark The estimate (7.2.4) is an equivalent condition for dissipativity of A, as discussed in Chapter 2. For insight into this, it is instructive to revisit the proof of Theorem 2.2.6. A proof of the equivalence can be found in Engel and Nagel [32], Proposition 3.23, pp. 81–2. In the Hilbert space case, the proof is much simpler, see, e.g., Sinha and Srivasta [91], Theorem 3.16, pp. 54–5.

Theorem 7.2.4 (Hille–Yosida–Ray Theorem) *A closed densely defined linear operator A is the generator of a positive contraction semigroup on $C_0(S)$ if and only if*

1. $Ran(\lambda I - A) = C_0(S)$ for all $\lambda > 0$.
2. A satisfies the positive maximum principle.

Proof. (Necessity) Suppose that $(T_t, t \geq 0)$ is a positive contraction semigroup on $C_0(S)$. Then (1) follows by the Lumer–Phillips theorem (Theorem 2.2.6). To establish the PMP (2), we assume that there exists $x_0 \in S$ for which $f(x_0) = \sup_{x \in S} f(x) \geq 0$. Define $f_+ = \max\{f, 0\}$. Then $f_+ \in C_0(S)$, $\|f_+\|_\infty = \|f\|_\infty$ and $f \leq f_+$. Then for all $t \geq 0$, using the positivity and contraction properties of T_t, we have

$$T_t f(x_0) \leq T_t f_+(x_0)$$
$$\leq \|T_t f_+\|_\infty$$
$$\leq \|f_+\|_\infty$$
$$= f(x_0).$$

Hence $Af(x_0) \leq 0$, as is required.

We are content here to just prove that A generates a contraction semigroup. In fact this follows from the Hille–Yosida theorem (Theorem 2.2.5) once we observe that the PMP implies (7.2.4) by Proposition 7.2.3, and so $\lambda I - A$ is injective from D_A to $C_0(S)$ for all $\lambda > 0$. Combining this with (1), we have that $\lambda I - A$ is bijective from D_A to $C_0(S)$ for all $\lambda > 0$. The resolvent estimate $\|R_\lambda\| \geq 1/\lambda$ for all $\lambda > 0$ then follows from (7.2.4). The proof of positivity is rather technical, and we refer the reader to Jacob [53], pp. 333–5 or Ethier and Kurtz [33], p. 166 for this. □

Our next task is to give a proof of the celebrated Courrège theorem that characterises a large class of linear operators that satisfy the PMP. For this we need to know a few things about distributions. So we begin with a brief survey of this topic, where we restrict the discussion to those results and ideas that we will need for our proof.

7.2.2 Crash Course on Distributions

We follow the superb account in Hörmander [49]. A distribution of order $N \in \mathbb{Z}_+$ is a linear functional $T : C_c^\infty(\mathbb{R}^d) \to \mathbb{R}$ such that for all compact $K \subseteq \mathbb{R}^d$ there exists $C_K \geq 0$ so that for all $\phi \in C_c^\infty(K)$,

$$|T(\phi)| \leq C_K \sum_{|\alpha| \leq N} ||D^\alpha \phi||_\infty. \tag{7.2.5}$$

We denote the set of all distributions of order N by $\mathcal{D}_N'(\mathbb{R}^d)$. It is clearly a real linear space. The space of all distributions of finite order is defined to be $\mathcal{D}'(\mathbb{R}^d) := \bigcup_{N \in \mathbb{Z}_+} \mathcal{D}_N'(\mathbb{R}^d)$.

It is customary to sometimes use the notation $\langle T, \phi \rangle$ instead of $T(\phi)$. Formally the product of a smooth function g and a distribution T of order N is a distribution of order N which we call T_g. This is defined as follows:

$$\langle T_g, \phi \rangle = \langle T, g\phi \rangle,$$

for all $\phi \in C_c^\infty(\mathbb{R}^d)$. Note that $g\phi \in C_c^\infty(\mathbb{R}^d)$, and so T_g is well-defined. As a special case, we use the notation T_q when q is the quadratic function $q(x) := |x|^2 = \sum_{i=1}^d x_i^2$ for all $x = (x_1, \ldots, x_d) \in \mathbb{R}^d$.

Let μ be a regular Borel measure on $(\mathbb{R}^d, \mathcal{B}(\mathbb{R}^d))$, and define $I_\mu : C_c^\infty(\mathbb{R}^d) \to \mathbb{R}$ by $I_\mu \phi = \int_{\mathbb{R}^d} \phi(x)\mu(dx)$. Then, for all compact $K \subseteq \mathbb{R}^d$, we have the inequality $|I_\mu \phi| \leq \mu(K)||\phi||_\infty$, and so I_ϕ is a distribution of order zero. In fact, every distribution of order zero is of this type. More generally we say that a distribution T is *positive* if $T\phi \geq 0$ whenever $\phi \geq 0$. Then, as is shown in Theorem 2.17 of Hörmander [49], p. 38, there exists a regular Borel measure μ so that $T = I_\mu$. This result is clearly very closely related to the Riesz lemma (see Theorem G.0.1).

The *support* of a distribution $T \in \mathcal{D}'(\mathbb{R}^d)$ is defined to be $\text{supp}(T) := \{y \in \mathbb{R}^d$; there is no open neighbourhood U_y of y for which $Tf = 0$ for all $\phi \in C_c^\infty(U_y)\}$. An important result for us will be Theorem 2.3.4 on pp. 46–7 of Hörmander [49], which states that if T is a distribution of order N for which $\text{supp}(T) = \{y\}$ for some $y \in \mathbb{R}^d$, then for all $\phi \in C_c^\infty(\mathbb{R}^d)$

$$T\phi = \sum_{|\alpha| \leq N} c_\alpha D^\alpha \phi(y), \tag{7.2.6}$$

where $c_\alpha \geq 0$ for each $1 \leq |\alpha| \leq N$.

7.2.3 The Courrège Theorem

In this section we aim to characterise all linear operators that satisfy the positive maximum principle (PMP). We will follow the succinct approach given by Walter Hoh in his Habilitationschrift [47] (see also Böttcher et al. [17], Proposition 2.27, pp. 52–4). First we need to extend the PMP to real-valued linear functionals T defined on the linear space $C_c^\infty(\mathbb{R}^d)$. We say that T satisfies the *positive maximum principle* (PMP), if, whenever $f \in C_c^\infty(\mathbb{R}^d)$ with $f(0) = \sup_{x \in \mathbb{R}^d} f(x) \geq 0$, then $Tf \leq 0$. The linear functional T is said to be *almost positive* if, whenever $f \in C_c^\infty(\mathbb{R}^d)$ with $f \geq 0$ and $f(0) = 0$, then $Tf \geq 0$.

Proposition 7.2.5 *If the linear functional* $T : C_c^\infty(\mathbb{R}^d) \to \mathbb{R}$ *satisfies the PMP, then it is almost positive.*

Proof. Suppose that $f \in C_c^\infty(\mathbb{R}^d)$ with $f \geq 0$ and $f(0) = 0$. Then $-f$ has a maximum at zero and so by the PMP, $-Tf = T(-f) \leq 0$. □

Proposition 7.2.6 *If the linear functional* $T : C_c^\infty(\mathbb{R}^d) \to \mathbb{R}$ *is almost positive, then it is a distribution of order two.*

Proof. Let $\theta \in C_c^\infty(\mathbb{R}^d)$ be such that $0 \leq \theta \leq 1$ and $\theta = 1$ in a neighbourhood of zero. Let $f \in C_c^\infty(\mathbb{R}^d)$ and define $\widetilde{f} \in C_c^\infty(\mathbb{R}^d)$ by

$$\widetilde{f}(x) = f(x) - f(0)\theta(x) - \sum_{i=1}^{d} \partial_i f(0)\theta(x)x_i,$$

for all $x = (x_1, \ldots, x_d) \in \mathbb{R}^d$. Now fix a compact set $K \subset \mathbb{R}^d$. Then it is easy to check that there exists $D_K > 0$ so that for all $x \in K, i, j = 1, \ldots, d$,

$$|\partial_i \partial_j \widetilde{f}(x)| \leq D_K \sum_{|\alpha| \leq 2} \|D^\alpha f\|_\infty.$$

Since $\widetilde{f}(0) = \partial_i \widetilde{f}(0) = 0$, a straightforward Taylor series expansion yields

$$|\widetilde{f}(x)| \leq \frac{1}{2} \sum_{i,j=1}^{d} \sup_{y \in K} |\partial_i \partial_j \widetilde{f}(y)|.|x_i x_j| \leq \frac{d}{2}|x|^2 D_K M_K,$$

where $M_K := \sum_{|\alpha| \leq 2} \|D^\alpha f\|_\infty$.

Next, choose a non-negative mapping $g_K \in C_c^\infty(\mathbb{R}^d)$ such that $g_K(x) = |x|^2$ for all $x \in K \cup \mathrm{supp}(\theta)$. Then we easily deduce that the mappings $G^\pm := \frac{d}{2} D_K M_K g_K \pm \widetilde{f} \geq 0$ with $G^\pm(0) = 0$. But T is almost positive and so $TG^\pm \geq 0$. From this we see that

$$|T\widetilde{f}| \leq \frac{d}{2} D_K M_K |Tg_K|.$$

Now return to the definition of $\tilde{f}$ to deduce that there exists $C_K > 0$ so that

$$|Tf| \leq C_K M_K,$$

and the required result follows. □

The next result is key as it establishes necessary conditions for a linear functional on $C_c^\infty(\mathbb{R}^d)$ to satisfy the positive maximum principle. Before you begin, it may be a good idea to re-familiarise yourself with the result of Theorem 3.5.14(3). There are intriguing similarities that will become clearer in the sequel.

Theorem 7.2.7 *Let T be a linear functional from $C_c^\infty(\mathbb{R}^d)$ to $\mathbb{R}$ which satisfies the PMP. Then there exists a unique quadruple (a, b, c, v) wherein a is a $d \times d$ non-negative definite symmetric matrix, b is a vector in $\mathbb{R}^d$, c is a non-negative constant and v is a Lévy measure on $\mathbb{R}^d$ so that for all $f \in C_c^\infty(\mathbb{R}^d)$*

$$Tf = -cf(0) + \sum_{i=1}^d b_i \partial_i f(0) + \sum_{i,j=1}^d a_{ij} \partial_i \partial_j f(0)$$

$$+ \int_{\mathbb{R}^d} \left(f(y) - f(0) - \sum_{i=1}^d y_i \partial_i f(0) I_{|y|<1}(y) \right) v(dy). \quad (7.2.7)$$

Proof. By Propositions 7.2.5 and 7.2.6, we see that T is a distribution of order two. Next we show that T_q is a positive distribution, where $q(x) = |x|^2$ for all $x \in \mathbb{R}^d$, as defined in section 6.2.2. To verify this, suppose that $f \in C_c^\infty(\mathbb{R}^d)$ with $f \geq 0$. Then $qf \geq 0$ with $qf(0) = 0$. Hence, since T is almost positive, we have $\langle T_q, f \rangle = \langle T, qf \rangle \geq 0$. Then, as explained in section 6.2.2, there exists a regular Borel measure μ on $\mathbb{R}^d$ so that $T_q f = \int_{\mathbb{R}^d} f(y)\mu(dy)$. Define a Borel measure v on $\mathbb{R}^d$ by

$$v(\{0\}) = 0 \text{ and } v(A) = \int_A \frac{1}{|y|^2} \mu(dy) \text{ if } A \in \mathcal{B}(\mathbb{R}^d \setminus \{0\}).$$

Then we have

$$Tf(x) = \int_{\mathbb{R}^d} f(x)v(dx),$$

for all $f \in C_c^\infty(\mathbb{R}^d)$ with $0 \notin \text{supp}(f)$.

We will show that v is a Lévy measure. First observe that

$$\int_{B_1(0)} |x|^2 v(dx) = \mu(B_1(0) \setminus \{0\}) < \infty,$$

since μ is regular. Now choose $g_i \in C_c^\infty(\mathbb{R}^d)$ with $0 \le g_i \le 1 (i = 1, 2)$ and with $\text{supp}(g_1) \subseteq \overline{B}_1(0)$ and $\text{supp}(g_2) \subseteq \overline{B}_1(0)^c$, with $g_1(0) = 1$. Then we clearly have

$$\sup_{x \in \mathbb{R}^d} (g_1(x) + g_2(x)) = g_1(0) = 1.$$

By the PMP, we have $T(g_1 + g_2) \le 0$, and so

$$\int_{\mathbb{R}^d} g_2(x)\nu(dx) = \langle T, g_2 \rangle \le -\langle T, g_1 \rangle.$$

Taking the supremum over all possible choices of g_2, we obtain $\nu(B_1(0)^c) < \infty$. So we have proved that ν is a Lévy measure. It is uniquely determined by this construction. Next we define the linear functionals $T_i \colon C_c^\infty(\mathbb{R}^d) \to \mathbb{R}$ for $i = 1, 2$ by

$$T_1 f = \int_{\mathbb{R}^d} \left(f(y) - f(0) - \sum_{i=1}^d y_i \partial_i f(0) \mathbf{1}_{|y|<1}(y) \right) \nu(dy),$$

and $T_2 = T - T_1$. It is immediate that T_1 satisfies the PMP and so is a distribution of order two. Hence we also have $T_2 \in \mathcal{D}_2'(\mathbb{R}^d)$. Furthermore, if 0 fails to be in the support of f, then $Tf = T_1 f = \int_{\mathbb{R}^d} f(x)\nu(dx)$, and so $T_2 f = 0$. Consequently $\text{supp}(T_2) \subseteq \{0\}$, and so by (7.2.6) we find that T_2 takes the form

$$T_2 = -cf(0) + \sum_{i=1}^d b_i \partial_i f(0) + \sum_{i,j=1}^d a_{ij} \partial_i \partial_j f(0),$$

and so T has the form (7.2.7), as required. It remains to establish that $a = (a_{ij})$ is a $d \times d$ non-negative definite symmetric matrix, and that c is a non-negative constant. We begin with c. Choose a sequence (g_n) in $C_c^\infty(\mathbb{R}^d)$ with $0 \le g_n \le 1$ so that $g_n = 1$ in a neighbourhood of zero for all $n \in \mathbb{N}$, and $g_n \uparrow 1$ as $n \to \infty$. By monotone convergence, we have

$$\lim_{n \to \infty} T_1 g_n = \lim_{n \to \infty} \int_{\mathbb{R}^d} (g_n(x) - 1)\nu(dx) = 0,$$

and so

$$\lim_{n \to \infty} T g_n = \lim_{n \to \infty} (T_1 g_n + T_2 g_n) = -\lim_{n \to \infty} c g_n(0) = -c,$$

but by the PMP, $T g_n \le 0$ for each $n \in \mathbb{N}$ and so $c > 0$, and is uniquely determined.

To deal with a, first observe that it is symmetric by construction. Now we choose a sequence (h_n) in $C_c^\infty(\mathbb{R}^d)$ with $0 \le h_n \le 1$, so that $h_n = 1$ in a neighbourhood of zero and $\text{supp}(h_n) \subseteq B_{1/n}(0)$ for all $n \in \mathbb{N}$. Since the linear functional T is almost positive, we have $T(h_n Q_y) \geq 0$ for all $n \in \mathbb{N}$ and all $y \in \mathbb{R}^d$, where Q_y denotes the quadratic function $Q_y(x) := \frac{1}{2}(x \cdot y)^2$ for all $x \in \mathbb{R}^d$. An easy calculation shows that

$$T_2(h_n Q_y) = \sum_{i,j=1}^d a_{ij} y_i y_j,$$

while $\lim_{n\to\infty} T_1(h_n Q_y) = \lim_{n\to\infty} \int_{\mathbb{R}^d} h_n(y) Q(y) \nu(dy) = 0$, by dominated convergence. Hence

$$\sum_{i,j=1}^d a_{ij} y_i y_j = \lim_{n\to\infty} T(h_n Q_y) \geq 0,$$

and so the matrix a is non-negative definite as required. It is also uniquely determined through the limit, and finally, since ν, b and a are now uniquely determined, it follows that b also is via the representation (7.2.7). $\qquad\square$

We also need to show sufficiency.

Proposition 7.2.8 *If T is a linear functional from $C_c^\infty(\mathbb{R}^d)$ to $\mathbb{R}$ which takes the form (7.2.7), then it satisfies the PMP.*

Proof. Let $f \in C_c(\mathbb{R}^d)$ with $f(0) = \sup_{x \in \mathbb{R}^d} f(x) \geq 0$. Then

$$Tf = \sum_{i,j=1}^d a_{ij} \partial_i \partial_j f(0) + \int_{\mathbb{R}^d} (f(x) - f(0)) \nu(dx).$$

Since $f(x) \leq f(0)$ for all $x \in \mathbb{R}^d$, it is clear that the second term is non-positive. For the first term, note that $\sum_{i,j=1}^d a_{ij} \partial_i \partial_j f(0) = \mathrm{tr}(aH)$, where $a = (a_{ij})$ and H is the Hessian matrix whose (i,j)th entry is $\partial_i \partial_j f(0)$. Now a is non-negative definite symmetric and so has a non-negative definite symmetric square root $a^{1/2}$. Similarly, since f takes its maximum at zero, $-H$ is non-negative definite symmetric and so has a non-negative definite symmetric square root $(-H)^{1/2}$. Then

$$\mathrm{tr}(aH) = -\mathrm{tr}(a^{1/2} a^{1/2} (-H)^{1/2} (-H)^{1/2})$$
$$= -\mathrm{tr}((-H)^{1/2} a^{1/2} a^{1/2} (-H)^{1/2})$$
$$= -\mathrm{tr}(C^* C) \leq 0,$$

where $C := a^{1/2}(-H)^{1/2}$, and the result follows. $\qquad\square$

Having established Theorem 7.2.7, we can obtain the important Courrège theorem as a corollary. In the following, $\mathbb{F}(\mathbb{R}^d)$ denotes the linear space of all mappings from $\mathbb{R}^d$ to $\mathbb{R}$.

Theorem 7.2.9 (Courrège theorem) *Let A be a linear operator from $C_c^\infty(\mathbb{R}^d)$ to $\mathbb{F}(\mathbb{R}^d)$. Then A satisfies the PMP if and only if there exists a unique quadruple $(a(\cdot), b(\cdot), c(\cdot), v(\cdot))$ wherein for all $x \in \mathbb{R}^d$, $a(x)$ is a $d \times d$ non-negative definite symmetric matrix, $b(x)$ is a vector in $\mathbb{R}^d$, $c(x)$ is a non-negative constant and $v(x, \cdot)$ is a Lévy measure on $\mathbb{R}^d$, so that for all $f \in C_c^\infty(\mathbb{R}^d)$*

$$Af(x) = -c(x)f(x) + \sum_{i=1}^d b_i(x)\partial_i f(x) + \sum_{i,j=1}^d a_{ij}(x)\partial_i\partial_j f(x)$$

$$+ \int_{\mathbb{R}^d} \left(f(x+y) - f(x) \right.$$

$$\left. - \sum_{i=1}^d y_i \partial_i f(x) \boldsymbol{I}_{|y|<1}(y) \right) v(x, dy). \qquad (7.2.8)$$

Proof. For each $x \in \mathbb{R}^d$, $f \in C_0(\mathbb{R}^d)$, let $\tau_x f(y) = f(y + x)$ for $y \in \mathbb{R}^d$ be the usual translation operator. So τ_x is an isometric isomorphism of $C_0(\mathbb{R}^d)$ which leaves $C_c^\infty(\mathbb{R}^d)$ invariant. We define a real-valued linear functional A_x on $C_c^\infty(\mathbb{R}^d)$ by $A_x f = A\tau_{-x} f(x) = (\tau_x A \tau_{-x} f)(0)$. It is clear that A satisfies the PMP if and only if the linear functional satisfies the PMP for every $x \in \mathbb{R}^d$. The result then follows from Theorem 7.2.7 and Proposition 7.2.8, and using $Af(x) = A_x \tau_x f$, for each $x \in \mathbb{R}^d$. □

The mapping $v \colon \mathbb{R}^d \times \mathcal{B}(\mathbb{R}^d) \to [0, \infty]$ which appears in (7.2.8) is called a *Lévy kernel*. We emphasise that the importance of the Courrège theorem is that it demonstrates the "Lévy–Khintchine" form that any generator A of a Feller semigroup must have if $C_c^\infty(\mathbb{R}^d) \subseteq D_A$, since by Proposition 7.2.2, all such operators satisfy the positive maximum principle. If we compare the result with that of Theorem 3.5.14(3), we see that if $c = 0$ (we will explain the role of c later), we can regard A as a field of Lévy operators indexed by $x \in \mathbb{R}^d$. This leads us to an intuitive view of the Feller process X as a field of Lévy processes. For an alternative proof of the Courrège theorem which does not involve the use of distributions, see Jacob [54], section 4.5 or the original paper by Courrège [23].

It is interesting to revisit convolution semigroups in the light of the Courrège theorem. So let $(\mu_t, t \geq 0)$ be a convolution semigroup on $\mathbb{R}^d$, and let $(T_t, t \geq 0)$ be the associated C_0–semigroup, so that, for all $t \geq 0, f \in C_0(\mathbb{R}^d), x \in \mathbb{R}^d$,

$$T_t f(x) = \int_{\mathbb{R}^d} f(x+y)\mu_t(dy).$$

It is easy to see that $T_t \tau_a = \tau_a T_t$ for all $a \in \mathbb{R}^d$. It follows that $\tau_a Af = A\tau_a f$ for all $f \in D_A$ for which $\tau_a f \in D_A$ for all $a \in \mathbb{R}^d$. In particular this last condition holds when $f \in C_c^\infty(\mathbb{R}^d)$. To verify the last identity, observe that for all $x \in \mathbb{R}^d$,

$$
\begin{aligned}
(\tau_a A f)(x) &= (Af)(x + a) \\
&= \lim_{t \to 0} \frac{1}{t} (T_t f(x + a) - f(x + a)) \\
&= \lim_{t \to 0} \frac{1}{t} (\tau_a T_t f(x) - \tau_a f(x)) \\
&= \lim_{t \to 0} \frac{1}{t} (T_t \tau_a f(x) - \tau_a f(x)) \\
&= A\tau_a f(x).
\end{aligned}
$$

Now since $(T_t, t \geq 0)$ is a Feller semigroup, A satisfies the PMP and so is of the form (7.2.8). So for all $x \in \mathbb{R}^d$, we have

$$
\begin{aligned}
(Af)(x) &= \tau_x A f(0) \\
&= A\tau_x f(0) \\
&= -c(0)f(x) + \sum_{i=1}^d b_i(0)\partial_i f(x) + \sum_{i,j=1}^d a_{ij}(0)\partial_i \partial_j f(x) \\
&\quad + \int_{\mathbb{R}^d} \left[f(x + y) - f(x) - \sum_{i=1}^d y_i \partial_i f(x) \mathbf{1}_{|y|<1} \right] \nu(0, dy).
\end{aligned}
$$

If we take $c(0) = 0$ and write $b = b(0), a = a(0)$ and $\nu = \nu(0, \cdot)$, we see that we have recaptured the representation (3.5.39). We will say more about the role of c in the next section. Finally we can at least formally derive the Lévy–Khintchine formula from (3.5.39) by computing $\eta(y) = (Ae_y)(0)$ for each $y \in \mathbb{R}^d$, where $e_y(x) = e^{ix \cdot y}$, for each $x \in \mathbb{R}^d$. Of course, $e_y \notin D_A$. This domain problem can be overcome, and to see how to do that, consult section 6 of the classic paper by Hunt [50]. An alternate and elegant proof of the Lévy–Khintchine formula, which bypasses (3.5.39) and proceeds directly from Theorem 7.2.7, may be found in Hoh [47], Theorem 2.13. The next result gives the key pseudo-differential operator representation.

Theorem 7.2.10 *If a linear operator $A \colon C_c^\infty(\mathbb{R}^d) \to \mathbb{F}(\mathbb{R}^d)$ satisfies the positive maximum principle, then it is a pseudo-differential operator with symbol $\eta \colon \mathbb{R}^d \times \mathbb{R}^d \to \mathbb{C}$ given for $x, y \in \mathbb{R}^d$ by*

$$
\eta(x, y) = c(x) + b(x) \cdot y + a(x)y \cdot y
$$
$$
+ \int_{\mathbb{R}^d} (e^{iu \cdot y} - 1 - iu \cdot y \mathbf{1}_{B_1(0)}(u))\nu(x, du), \qquad (7.2.9)
$$

i.e., for all $f \in C_c^\infty(\mathbb{R}^d)$,

$$Af(x) = \frac{1}{(2\pi)^{d/2}} \int_{\mathbb{R}^d} \eta(x, y) e^{ix \cdot y} \widehat{f}(y) dy.$$

Proof. This follows from similar arguments to those used in the proof of Theorem 3.5.14. We begin by using Fourier inversion to write $f(x) = \frac{1}{(2\pi)^{d/2}} \int_{\mathbb{R}^d} e^{ix \cdot y} \widehat{f}(y) dy$ and then compute $Af(x)$, using Theorem 7.2.8. Details are left to the reader. $\square$

Its worth pointing out that one of the distinguishing features of the way in which pseudo-differential operators appear in this context, as opposed to their standard use in partial differential equations, is the lack of smoothness of the symbol $x \to \eta(x, \cdot)$. In general, we can only expect this mapping to be measurable.

We now know a great deal about linear operators that satisfy the positive maximum principle. But in order for such an operator to generate a Feller semigroup, it must at the very least map $C_c^\infty(\mathbb{R}^d)$ to $C_0(\mathbb{R}^d)$. Sufficient conditions for this to hold are given in the next result. Finer results on continuity can be found in the original paper Courrège [23]; see also Lemma 4.5.16 in Jacob [54], pp. 349–51. First we need some definitions. We say that a Lévy kernel ν has a *Lévy density* with respect to some σ–finite Borel measure ρ on $\mathbb{R}^d$ (which might, e.g., be Lebesgue measure) if there is a measurable function $h_\nu \colon \mathbb{R}^d \times \mathbb{R}^d \to [0, \infty)$ such that for all $x \in \mathbb{R}^d$, $h_\nu(x, \cdot)$ is the Radon–Nikodym derivative of the Lévy measure $\nu(x, \cdot)$ with respect to ρ; i.e., for all $B \in \mathcal{B}(\mathbb{R}^d)$,

$$\nu(x, B) = \int_B h_\nu(x, y) \rho(dy).$$

Such a Lévy kernel is said to be *Lévy continuous* if, for all $x \in \mathbb{R}^d$ and all sequences (x_n) that converge to x, we have

$$\lim_{n \to \infty} \int_{\mathbb{R}^d} (|y|^2 \wedge 1) |h_\nu(x, y) - h_\nu(x_n, dy)| \rho(dy) = 0.$$

Theorem 7.2.11 *If a linear operator $A \colon C_c^\infty(\mathbb{R}^d) \to \mathbb{F}(\mathbb{R}^d)$ satisfies the positive maximum principle, then for each $f \in C_c^\infty(\mathbb{R}^d)$,*

1. *Af is continuous if, in (7.2.8), c, b_i and a_{jk} are continuous and the Lévy kernel ν is Lévy continuous.*
2. *Af vanishes at infinity if, in (7.2.8),*

$$\lim_{|x| \to \infty} \int_{\mathbb{R}^d} (|y|^2 \wedge 1) \nu(x, dy) = 0.$$

Proof. It is sufficient to consider the case where $b_i = a_{jk} = c = 0$, for all $i, j, k = 1, \ldots, d$. We write $A = A_1 + A_2$, where for each $f \in C_c^\infty(\mathbb{R}^d)$, $x \in \mathbb{R}^d$,

$$A_1 f(x) = \int_{B_1(0)} [f(x+y) - f(x) - y \cdot \nabla f(x)] v(x, dy),$$

$$A_2 f(x) = \int_{B_1(0)^c} [f(x+y) - f(x)] v(x, dy).$$

1. Let (x_n) be a sequence converging to $x \in \mathbb{R}^d$. Then for each $n \in \mathbb{N}$,

$$A_2 f(x) - A_2 f(x_n)$$

$$= \int_{B_1(0)^c} [f(x+y) - f(x) - f(x_n + y) + f(x_n)] v(x, dy)$$

$$+ \int_{B_1(0)^c} [f(x_n + y) - f(x_n)](h_v(x, y) - h_v(x_n, y)) \rho(dy).$$

The first term tends to zero, as $n \to \infty$, by dominated convergence. The absolute value of the second term is majorised by

$$2\|f\|_\infty \int_{B_1(0)^c} |h_v(x, y) - h_v(x_n, y)| \rho(dy) \to 0 \text{ as } n \to \infty,$$

by definition of Lévy continuity. Similarly we obtain

$$A_1 f(x) - A_1 f(x_n)$$

$$= \int_{B_1(0)} [f(x+y) - f(x) - y \cdot \nabla f(x) - f(x_n + y) + f(x_n)$$

$$+ y \cdot \nabla f(x_n)] v(x, dy)$$

$$+ \int_{B_1(0)} [f(x_n + y) - f(x_n) - y \cdot \nabla f(x_n)](h_v(x, y) - h_v(x_n, y)) \rho(dy).$$

Using a standard Taylor series expansion, the absolute value of the second term is majorised by

$$d \max_{1 \leq i, j \leq d} \|\partial_i \partial_j f\|_\infty \int_{B_1(0)} |y|^2 |h_v(x, y) - h_v(x_n, y)| \rho(dy) \to 0 \text{ as } n \to \infty,$$

by definition of Lévy continuity. The first term is dealt with by a Taylor expansion and dominated convergence.

2. The result follows easily from the estimates

$$|A_1 f(x)| \leq d \max_{1 \leq i, j \leq d} \|\partial_i \partial_j f\|_\infty \int_{B_1(0)} |y|^2 v(x, dy)$$

and

$$|A_2 f(x)| \leq 2\|f\|_\infty v(x, B_1(0)^c). \qquad \square$$

If all of the conditions of Theorem 7.2.11 are satisfied, then A maps $C_c^\infty(\mathbb{R}^d)$ to $C_0(\mathbb{R}^d)$ and may be considered as a linear operator on $C_0(\mathbb{R}^d)$ with domain $C_c^\infty(\mathbb{R}^d)$. More generally, if A is a closed linear operator in $C_0(\mathbb{R}^d)$ with $C_c^\infty(\mathbb{R}^d) \subseteq D_A$ and A satisfies the positive maximum principle, then A is the generator of a positive contraction semigroup if and only if it satisfies the "elliptic" condition (1) of the Hille–Yosida–Ray Theorem 7.2.4. For a review of some direct approaches to solving the problem of finding conditions under which (1) holds for a linear operator that satisfies the PMP, see the discussion and references in Böttcher et al., pp. 70–4. In particular, Jacob [52] uses the pseudo-differential operator representation of Theorem 7.2.10 and imposes conditions on the symbol which ensure that the condition is satisfied. An alternative approach, which is a key theme of Ethier and Kurtz [33], is to use probabilistic arguments to show that A is associated to a Feller process by solving the *martingale problem* (see below). Then A will in fact generate a Feller semigroup. This theme is picked up in Hoh [47, 48] for pseudo-differential operators. We recommend Böttcher et al. [17] for a comprehensive review of the construction of Feller processes from suitable symbols of pseudo-differential operators, and discussion of how properties of the symbol encode those of the process.

The power of pseudo-differential operator techniques has been a key theme so far in twenty-first-century work on Feller processes. The case where $\nu = 0$ in (7.2.8) is of great importance, as that reconnects us with the theme of partial differential equations. Much of Chapter 2 of Jacob [55] is devoted to finding conditions on the coefficients $c(\cdot)$, $b(\cdot)$ and $a(\cdot)$ so that such an A extends to be the generator of a positive contraction semigroup. In particular, Theorem 2.1.43 therein proves that if c is a non-positive function, and the operator is uniformly elliptic with a taking values in the symmetric matrices, and furthermore that $c \in C_b^1(\mathbb{R}^d)$, $b_i \in C^2(\mathbb{R}^d)$ and $a_{jk} \in C_b^3(\mathbb{R}^d)$ for all $i, j, k = 1, \ldots d$, then A does indeed have such an extension. Ideally we would like to have a refinement of the Hille–Yosida–Ray theorem that gives necessary and sufficient conditions for a linear operator to generate a Feller semigroup (in the sense given herein). That seems to be a very hard problem.

7.3 The Martingale Problem

Having mentioned the martingale problem, we should say just a little about it. Let $(\Omega, \mathcal{F}, P)$ be a probability space equipped with a filtration $(\mathcal{F}_t, t \geq 0)$ of sub-σ-algebras of $\mathcal{F}$. A process $(Y(t), t \geq 0)$ defined on our probability space, and adapted to the filtration, is said to be a *martingale* if $\mathbb{E}(|Y(t)|) < \infty$ for all $t \geq 0$, and for all $0 \leq s \leq t < \infty$,

$$\mathbb{E}(Y(t)|\mathcal{F}_s) = Y(s).$$

A martingale is *centered* if $\mathbb{E}(Y(t)) = 0$ for all $t \geq 0$.

There are many important examples of martingales in the theory of stochastic processes. Brownian motion is probably the best known of these. A Lévy process $(L(t), t \geq 0)$ with characteristics (b, a, ν) is a martingale if and only if $\int_{|x|>1} |x|\nu(dx) < \infty$, and $b - \int_{|x|>1} x\nu(dx) = 0$ (see, e.g., Applebaum [6], p. 133). Moreover the martingale property is inherited by stochastic integrals of the form $\int_0^t F(s)dM(s)$, where M is a square-integrable continuous martingale, and the adapted process F is sufficiently regular. For more about martingales, see Appendix E.

If the reader knows a little bit about stochastic differential equations, the next example is good motivation for the result that follows. Consider the linear operator A defined on $C_c^\infty(\mathbb{R}^d)$ by

$$Af(x) = \sum_{i=1}^d b_i(x)\partial_i f(x) + \sum_{i,j=1}^d a_{ij}(x)\partial_i \partial_j f(x),$$

for all $f \in C_c^\infty(\mathbb{R}^d)$, $x \in \mathbb{R}^d$, and assume that the coefficients b and a are sufficiently regular so that A generates a Feller semigroup in $C_0(\mathbb{R}^d)$. Assume further that $a(\cdot)$ has a square-root $\sigma(\cdot)$ so that $\sigma(x)\sigma(x)^T = a(x)$ for all $x \in \mathbb{R}^d$, and that the stochastic differential equation:

$$dY_i(t) = b_i(Y(t)) + \sum_{j=1}^d \sigma_{ij}(Y(t))dB_i(t),$$

for all $i = 1, \ldots, d$, with initial condition $Y(0) = Y_0$ (a.s.) has a unique solution. Then by Itô's formula, for all $f \in C_c^\infty(\mathbb{R}^d)$, $t \geq 0$,

$$f(Y(t)) - f(Y_0) - \int_0^t (Af)(Y(s))ds = \int_0^t \sum_{i,j=1}^d \sigma_{ij}(Y(s))\partial_j f(Y(s))dB_i(s),$$

so that the process whose value at t is $f(Y(t)) - f(Y_0) - \int_0^t (Af)(Y(s))ds$, is a centred martingale.[3]

---

[3] Provided σ is sufficiently regular. Otherwise it will be a local martingale, but this is not the place to pursue that theme further.

Proposition 7.3.12 *If $X = (X(t), t \geq 0)$ is a Feller process with generator A, then for all $f \in D_A$, $M_f = (M_f(t), t \geq 0)$ is a martingale, where for all $t \geq 0$,*

$$M_f(t) := f(X(t)) - f(X(0)) - \int_0^t (Af)(X(s))ds. \qquad (7.3.10)$$

Furthermore, if we fix $x \in \mathbb{R}^d$ so that $X(0) = x$ (a.s.), then M_f is centred.

Proof. Let $(T_t, t \geq 0)$ be the Feller semigroup generated by A and fix $t > 0$. It is easy to verify the integrability condition, e.g., we have the straightforward estimate $\mathbb{E}\left(\left|\left(\int_0^t (Af)(X(s))ds\right)\right|\right) \leq t\|Af\|_\infty$. To establish the martingale property, if $0 \leq s \leq t$, we have,[4] by the Markov property (7.1.3) and Lemma 1.3.14(2)

$$\mathbb{E}(M_f(t) - M_f(s)|\mathcal{F}_s)$$

$$= \mathbb{E}(f(X(t))|\mathcal{F}_s) - f(X(s)) - \int_s^t \mathbb{E}((Af)(X(u))|\mathcal{F}_s)du$$

$$= T_{t-s}f(X(s)) - f(X(s)) - \int_s^t (T_{u-s}Af)(X(s))du$$

$$= T_{t-s}f(X(s)) - f(X(s)) - \int_s^t \frac{d}{du}(T_{u-s}f)(X(s))du$$

$$= 0,$$

and the first result follows. For the second, if $X(0) = x$ (a.s.), then by Fubini's theorem,

$$\mathbb{E}(M_f(t)) = T_t f(x) - f(x) - \int_0^t T_s Af(x)ds = 0,$$

using (1.3.9). □

For the martingale problem, we are given a linear operator A on some domain $D_A \subseteq C_0(\mathbb{R}^d)$ and the goal is to construct a Feller process X on some probability space such that, with $X(0) = x$ (a.s.), the prescription (7.3.10) yields a centred martingale. Then we get a Feller semigroup as a byproduct of the procedure. For more about this, a standard reference is Ethier and Kurtz [33]. For recent developments, in which the data with which we start is the symbol of a suitable pseudo-differential operator, see, e.g., Hoh [47, 48], Chapter 4 of Jacob [56] and Chapter 5 of Kolokoltsov [60].

---

[4] Using a conditional Fubini theorem, as in Applebaum [6], Theorem 1.1.8, p. 13.

7.3.1 Sub-Feller Semigroups

A contraction semigroup on $C_0(S)$ is said to be *sub-Feller* if $f \in C_0(S)$ with $0 \leq f \leq 1$ implies that $0 \leq T_t f \leq 1$ for all $t \geq 0$. Note that every Feller semigroup is sub-Feller, as is easily seen from the integral representation

$$T_t f(x) = \int_S f(y) p_t(x, dy),$$

where $x \in S$. On the other hand a sub-Feller semigroup is Feller if and only if it is conservative. To see this it is sufficient to prove that a sub-Feller semigroup is positive, and to verify that we choose a non-zero, non-negative $f \in C_0(S)$. Then $0 \leq f/\|f\|_\infty \leq 1$ and so $1/\|f\|_\infty T_t f \geq 0$. Hence $T_t f \geq 0$ as required.

It can be shown that every sub-Feller semigroup is the transition semigroup of a Markov process that takes values, not in S but in its one-point compactification. Defining transition probabilities as above, we have $p_t(x, S) \leq 1$ with equality if and only if the semigroup is Feller. The argument in the proof of Proposition 7.2.2 shows that the generator A of a sub-Feller semigroup satisfies the PMP. Note, however, that if the semigroup is conservative and $1 \in D_A$, where A is now considered as an extended operator, acting in $UC_b(S)$, we must have $A1 = 0$. However, (formally speaking) (7.2.8) yields $A1(x) = -c(x)$ for each $x \in \mathbb{R}^d$ and so the function c appearing there is the obstacle to conservativity, at the infinitesimal level. The real story is more complicated than this, for how do we know that the form (7.2.8) still holds for the extended operator on its domain in $UC_b(S)$? To learn more about this, see Bottcher et al. [17] (Lemma 2.32 and Theorem 2.33 on p. 56) and references therein. The use of pseudo-differential operator techniques, based on the representation for the symbol of A given in Theorem 7.2.10, appear to be essential to obtain these results.

Note that the function c may be interpreted as the rate at which paths of the process are "killed" in S and sent to the point at infinity. The Feynman–Kac formula can be used to gain insight into this. For details, see Applebaum [6], section 6.7.2, pp. 405–6.

The above chain of ideas is very simple when we have a convolution semigroup $(\mu_t, t \geq 0)$ and we take $c > 0$ to be constant. For each $t \geq 0$, define $\widetilde{\mu}_t := e^{-ct}\mu_t$. Then $(\widetilde{\mu}_t, t \geq 0)$ is a convolution semigroup of sub-probability measures in that $\widetilde{\mu}_t(\mathbb{R}^d) = e^{-ct} < 1$ when $t > 0$. In this case the sub-Feller semigroup $(\widetilde{T}_t, t \geq 0)$ on $C_0(\mathbb{R}^d)$ is given for all $f \in C_0(\mathbb{R}^d), x \in \mathbb{R}^d$, $t \geq 0$ by

$$\widetilde{T}_t f(x) = \int_{\mathbb{R}^d} f(x + y)\widetilde{\mu}_t(dy) = e^{-ct} \int_{\mathbb{R}^d} f(x + y)\mu_t(dy).$$

In this case we have $\widetilde{T}_t = e^{-ct} T_t$ for all $t \geq 0$ where $(T_t, t \geq 0)$ is the usual C_0-semigroup (with generator A) associated to the convolution semigroup. Then by Problem 1.7, the action of the generator of our sub-Feller semigroup is $\widetilde{A} = -cI + A$ on D_A, and we can establish the form (7.2.8) from (3.5.39).

8

Semigroups and Dynamics

8.1 Invariant Measures and Entropy

8.1.1 Invariant Measures

Let $(T_t, t \geq 0)$ be a Markov semigroup on a locally compact space[1] S. We say that a Borel measure μ is an *invariant measure* for $(T_t, t \geq 0)$ if

$$\int_{\mathbb{R}^d} T_t f(x)\mu(dx) = \int_{\mathbb{R}^d} f(x)\mu(dx), \tag{8.1.1}$$

for all $t \geq 0$, $f \in B_b^+(\mathbb{R}^d)$; e.g., if $(T_t, t \geq 0)$ is the transition semigroup for a Lévy process associated to a convolution semigroup $(\rho_t, t \geq 0)$ on $\mathbb{R}^d$, then Lebesgue measure is an invariant measure, as by Fubini's theorem

$$\int_{\mathbb{R}^d} T_t f(x)dx = \int_{\mathbb{R}^d} \int_{\mathbb{R}^d} f(x+y)\rho_t(dy)dx = \int_{\mathbb{R}^d} f(x)dx.$$

Proposition 8.1.1 *If $(T_t, t \geq 0)$ is a Markov semigroup on S having transition probabilities p_t for $t \geq 0$, then μ is an invariant measure if and only if*

$$\int_{\mathbb{R}^d} p_t(x, A)\mu(dx) = \mu(A), \tag{8.1.2}$$

for all $A \in \mathcal{B}(\mathbb{R}^d)$.

Proof. For necessity, just take $f = \mathbf{1}_A$ in (8.1.1). For sufficiency, since (8.1.2) is (8.1.1) for indicator functions, we can extend by linearity to non-negative simple functions, and then use monotone convergence to obtain the result for $f \in B_b^+(S)$. $\square$

---

[1] Once again, feel free to take $S \subseteq \mathbb{R}^d$ throughout this chapter if you have not yet met this notion.

For many applications, we may require μ to be an invariant probability measure; e.g., for convolution semigroups on the circle, normalised Lebesgue measure $dx/(2\pi)$ is an invariant probability measure. A more sophisticated example can be found on pp. 404–5 of Applebaum [6] where we consider the Mehler semigroup associated to the Ornstein–Uhlenbeck process driven by a Brownian motion. In this case, there is a Gaussian invariant measure.

The following proposition is a useful consequence of the existence of an invariant measure.

Proposition 8.1.2 *If $(T_t, t \geq 0)$ is a Markov semigroup on S having a regular invariant measure μ, then for each $t \geq 0$, the action of T_t on $C_c(S)$ extends to a contraction on $L^p(S, \mu)$ for all $1 \leq p < \infty$, with (S1) and (S2) being satisfied on those spaces.*

Proof. Using Jensen's inequality, Fubini's theorem and (8.1.1), for all $t \geq 0$, $f \in C_c(S)$, we have

$$
\begin{aligned}
\|T_t f\|_p^p &= \int_S \left| \int_S f(y) p_t(x, dy) \right|^p \mu(dx) \\
&\leq \int_S \int_S |f(y)|^p p_t(x, dy) \mu(dx) \\
&= \int_S |f(y)|^p \mu(dy) = \|f\|_p^p.
\end{aligned}
$$

The extension to L^p is then an easy density argument. Verification of (S1) and (S2) is left to the reader.   □

If the transition probabilities $p_t(x, \cdot)$ are absolutely continuous with respect to μ, for all $x \in S, t > 0$, the corresponding Radon–Nikodym derivatives $k_t(x, \cdot)$ are called *transition densities* for the process.

In general, if $(T_t, t \geq 0)$ is a self-adjoint semigroup in $L^2(S, \mu)$, then it is called a *symmetric Markov semigroup*. The symmetric diffusion semigroups that we investigated in section 5.2 are special cases of this, where μ was taken to be Lebesgue measure. Another class of examples is semigroups associated to symmetric convolution semigroups. Symmetric Markov semigroups are a major focus of the theory of *Dirichlet forms*; see Fukushima et al. [38]. If a symmetric Markov process has transition densities, then it is easily verified that these are symmetric almost everywhere, i.e.,

$$
k_t(x, y) = k_t(y, x),
$$

for all $(x, y) \in (S \times S) \setminus E$ where $E \in \mathcal{B}(S \times S)$ with $(\mu \times \mu)(E) = 0$.

Invariant measures are the starting point for studying mixing properties of semigroups. For example, we say that a Feller semigroup $(T_t, t \geq 0)$ equipped with an invariant probability measure μ is *ergodic* if, whenever $A \in \mathcal{B}(\mathbb{R}^d)$ with $T_t 1_A = 1_A$ (μ a.e.) for all $t > 0$, then $\mu(A) = 0$ or $\mu(A) = 1$. One of the most important consequences of ergodicity is the fact that "time averages" equal "space averages". This was in fact the motivation from physics that led to the development of the mathematical ideas. We state this property as follows: if $X = (X(t), t \geq 0)$ is the Feller process whose semigroup is $(T_t, t \geq 0)$, then ergodicity implies

$$\lim_{T \to \infty} \frac{1}{T} \int_0^T f(X(s))ds = \int_{\mathbb{R}^d} f(x)\mu(dx),$$

for all $f \in B_b(\mathbb{R}^d)$. A good reference for this theory is da Prato and Zabczyk [24]. We mention one other interesting result that can be found therein. It is easy to see that the set of all invariant measures for a Feller semigroup is convex. It turns out that the ergodic measures (i.e., those for which the semigroup is ergodic) are the extreme points of this convex set. It follows that if an invariant measure is unique, then it is ergodic.

8.1.2 Entropy

Let $(T_t, t \geq 0)$ be a Feller semigroup with an invariant measure μ. Assume that $f \in L^1_+(\mathbb{R}^d, \mu)$ is such that the quantity

$$H_\mu(f) := -\int_{\mathbb{R}^d} f(x) \log(f(x))\mu(dx)$$

is finite. In that case we call H_μ the entropy of f. The famous *second law of thermodynamics* states that the entropy of a closed system can never decrease. Here is a mathematical formulation of this result which is due to Yosida [101], p. 392. Here we are considering the semigroup as modelling the evolution of a system in time. The next section will explain how this can arise from semidynamical or dynamical systems.

Theorem 8.1.3 *If $(T_t, t \geq 0)$ is a Feller semigroup with invariant probability measure μ, then for all $f \in L^1_+(\mathbb{R}^d, \mu)$ with $H_\mu(f) < \infty$ and $t \geq 0$,*

$$H_\mu(f) \leq H_\mu(T_t f).$$

Proof. Let $h(x) := -x \log(x)$ for $x > 0$. Then h is twice differentiable with $h''(x) = -1/x < 0$, and so h is a concave function. Hence we may apply Jensen's inequality and (8.1.1) to obtain

$$H_\mu(T_t f) = \int_{\mathbb{R}^d} h(T_t f(x)) \mu(dx)$$

$$= \int_{\mathbb{R}^d} h\left(\int_{\mathbb{R}^d} f(y) p_t(x, dy)\right) \mu(dx)$$

$$\geq \int_{\mathbb{R}^d} \int_{\mathbb{R}^d} h(f(y)) p_t(x, dy) \mu(dx)$$

$$= \int_{\mathbb{R}^d} T_t h(f(x)) \mu(dx)$$

$$= \int_{\mathbb{R}^d} h(f(x)) \mu(dx)$$

$$= H_\mu(f). \qquad\qquad \square$$

8.2 Semidynamical Systems

8.2.1 Koopmanism

Let S be a locally compact space. A *semidynamical system* is a family $(\alpha_t, t \geq 0)$ of mappings from S to itself, satisfying the following conditions:

(i) $\alpha_0 = \text{id}$.
(ii) $\alpha_s \circ \alpha_t = \alpha_{s+t}$, for all $s, t \geq 0$.
(iii) The mapping $(t, x) \to \alpha_t(x)$ is continuous from $[0, \infty) \times S$ to S.

A *dynamical system* $(\alpha_t, t \in \mathbb{R})$ satisfies (i) and modified versions of (ii) and (iii), with $[0, \infty)$ replaced by $\mathbb{R}$ and

(iv) For each $t \in \mathbb{R}$, α_t is invertible with $\alpha_t^{-1} = \alpha_{-t}$.

So a semidynamical system is a continuous semigroup of mappings of S to itself, while a dynamical system is a group of such mappings. Our aim in this section will be to look at how these lead to associated C_0-semigroups and groups on function spaces, as was briefly alluded to in Example 1.2.7. We will mainly deal with dynamical systems, as the theory is more straightforward. First let us explore invariant measures in this context. We say that a finite measure μ on $(S, \mathcal{B}(S))$ is an *invariant measure* for the semidynamical system $(\alpha_t, t \geq 0)$ if $\mu(\alpha_t^{-1}(A)) = \mu(A)$, for all $t \geq 0$, $A \in \mathcal{B}(S)$, and we say that μ is *quasi-invariant* if

$$\mu(A) = 0 \Rightarrow \mu(\alpha_t^{-1}(A)) = 0.$$

In the quasi-invariant case, each of the measures $\mu_t := \mu \circ \alpha_t^{-1}$ is absolutely continuous with respect to μ, and the Radon–Nikodym theorem[2] then yields the existence of a family of densities $g_t^{\mu} := \frac{d\mu_t}{d\mu} \in L^1(S, \mu)$. For a dynamical system, these densities satisfy a *cocycle condition* that will be central in the discussion that follows:

Theorem 8.2.4 *If μ is a quasi-invariant measure for the dynamical system $(\alpha_t, t \in \mathbb{R})$, then for all $s, t \in \mathbb{R}$,*

$$g_{s+t}^{\mu} = (g_s^{\mu} \circ \alpha_{-t})g_t^{\mu} \quad (a.e.\ \mu). \tag{8.2.3}$$

Proof. For all $A \in \mathcal{B}(S)$,

$$\mu(\alpha_{s+t}^{-1}(A)) = \int_A g_{s+t}^{\mu}(x)\mu(dx).$$

We also have

$$\mu(\alpha_{s+t}^{-1}(A)) = \mu(\alpha_s^{-1}(\alpha_t^{-1}(A)))$$
$$= \int_{\alpha_t^{-1}(A)} g_s^{\mu}(x)\mu(dx)$$
$$= \int_A g_s^{\mu}(\alpha_t^{-1}(x))\mu_t(dx)$$
$$= \int_A g_s^{\mu}(\alpha_t^{-1}(x))g_t^{\mu}(x)\mu(dx),$$

and the result follows by the uniqueness of the Radon–Nikodym derivative (see Appendix E). $\qquad\square$

We say that the dynamical system $(\alpha_t, t \geq 0)$ is *quasi-invariant* if each $\mu \circ \alpha_t^{-1}$ is quasi-invariant, and *continuously quasi-invariant* if g_t^{μ} is continuous for all $t \in \mathbb{R}$ and $\lim_{t \to 0} g_t^{\mu}(x) = 1$ for all $x \in S$. In the next subsection, we will need the notion of a *differentiably quasi-invariant* dynamical system wherein we assume that the mapping $t \to g_t^{\mu}(x)$ from $[0, \infty)$ to $[0, \infty)$ is differentiable at $t = 0$ for all $x \in S$, and that the mapping $x \to \frac{d}{dt}g_t^{\mu}(x)\big|_{t=0}$ is a continuous function on S.

For all $f \in C_c(S), 1 \leq p < \infty, t \in \mathbb{R}, x \in S$, define $P_t: C_c(S) \to C_c(S)$ by

$$P_t f(x) = f(\alpha_{-t}(x))(g_t^{\mu}(x))^{\frac{1}{p}}. \tag{8.2.4}$$

[2] See Appendix E for a proof of this result.

Then P_t is clearly linear and

$$\|P_t f\|_p^p = \int_S |f(\alpha_{-t}(x))|^p g_t^\mu(x) \mu(dx)$$

$$= \int_S |f(\alpha_{-t}(x))|^p \mu_t(dx)$$

$$= \int_S |f(x)|^p \mu(dx),$$

and so for all $t \in \mathbb{R}$, P_t extends to an isometry on $L^p(S, \mu)$. We call the isometry P_t a *Perron–Frobenius operator*.

Theorem 8.2.5 *If the dynamical system* $(\alpha_t, t \in \mathbb{R})$ *is continuously quasi-invariant, then* $(P_t, t \geq 0)$ *is a C_0-group of isometries of $L^p(S, \mu)$.*

Proof. $P_0 = I$ is obvious. For $f \in C_c(S)$, $s, t \geq 0$, $x \in S$, we have

$$P_s(P_t f)(x) = P_t f(\alpha_{-s}(x))(g_s^\mu(x))^{\frac{1}{p}}$$

$$= f(\alpha_{-s-t}(x))(g_t^\mu(\alpha_{-s}(x)))^{\frac{1}{p}} g_s^\mu(x)^{\frac{1}{p}}$$

$$= f(\alpha_{-s-t}(x))(g_{s+t}^\mu(x))^{\frac{1}{p}} = P_{s+t} f(x),$$

by the cocycle property (Theorem 8.2.4), and the result extends to $f \in L^p(S, \mu)$ by density.

For the strong continuity, we have by Minkowski's inequality

$$\|P_t f - f\|_p = \left(\int_S |f(\alpha_{-t}(x))(g_t^\mu(x))^{\frac{1}{p}} - f(x)|^p \mu(dx) \right)^{\frac{1}{p}}$$

$$\leq \left(\int_S |f(\alpha_{-t}(x))((g_t^\mu(x))^{\frac{1}{p}} - 1)|^p \mu(dx) \right)^{\frac{1}{p}}$$

$$+ \left(\int_S |f(\alpha_{-t}(x)) - f(x)|^p \mu(dx) \right)^{\frac{1}{p}}.$$

Both integrals converge to zero as $t \to 0$, by using dominated convergence, and continuity of the dynamical system in the second integral, and continuous quasi-invariance in the first. The limiting relation extends to $f \in L^p(S, \mu)$ by density, and a standard $\epsilon/3$ argument. $\square$

Next we establish a *consistency property* for the Radon–Nikodym derivatives.

Proposition 8.2.6 (Consistency) *If μ is a quasi-invariant measure for the dynamical system* $(\alpha_t, t \in \mathbb{R})$, *then for all $t \in \mathbb{R}$, $x \in S$*

$$g_t^\mu(\alpha_t(x)) g_{-t}^\mu(x) = 1 \quad (a.e. \ \mu). \tag{8.2.5}$$

Proof. For all $f \in C_c(S)$,

$$\int_S f(x)\mu(dx) = \int_S f(\alpha_{-t}(\alpha_t(x)))\mu(dx)$$

$$= \int_S f(\alpha_{-t}(x))g_t^\mu(x)\mu(dx)$$

$$= \int_S f(x)g_t^\mu(\alpha_t(x))g_{-t}^\mu(x)\mu(dx),$$

and the result follows from uniqueness in the Radon–Nikodym theorem. $\square$

Having fixed $1 < p < \infty$, choose $q = p/(p-1)$ so that $L^q(S, \mu)$ is the Banach space dual to $L^p(S, \mu)$. For all $f \in C_c(S), t \in \mathbb{R}, x \in S$, define $U_t: C_c(S) \to C_c(S)$ by

$$U_t f(x) = f(\alpha_t(x))(g_{-t}^\mu(x))^{\frac{1}{q}}. \tag{8.2.6}$$

We call U_t a *Koopman operator*. Of course, we may write $U_t = P_{-t}$, where $(P(t), t \in \mathbb{R})$ is the corresponding group of Perron–Frobenius operators acting in $L^q(S, \mu)$. We then deduce immediately from Theorem 8.2.5 that $(U_t, t \in \mathbb{R})$ is a group of isometries of $L^q(S, \mu)$.

Theorem 8.2.7 *The Koopman operators $(U_t, t \in \mathbb{R})$ are the adjoint group to the C_0-group of Perron–Frobenius operators $(P_t, t \in \mathbb{R})$.*

Proof. Fix $t \in \mathbb{R}$, and let $f, h \in C_c(S)$, then by consistency (Proposition 8.2.6)

$$\langle P_t f, h \rangle = \int_S f(\alpha_{-t}(x))(g_t^\mu(x))^{\frac{1}{p}} h(x)\mu(dx)$$

$$= \int_S f(x)(g_t^\mu(\alpha_t(x)))^{\frac{1}{p}} g_{-t}^\mu(x) h(\alpha_t(x))\mu(dx)$$

$$= \int_S f(x)(g_t^\mu(\alpha_t(x)))^{\frac{1}{p}} (g_{-t}^\mu(x))^{\frac{1}{p}} (g_{-t}^\mu(x))^{\frac{1}{q}} h(\alpha_t(x))\mu(dx)$$

$$= \langle f, U_t h \rangle,$$

and the result follows, by a density argument. $\square$

A straightforward corollary is that $(P_t, t \in \mathbb{R})$ is a unitary group when $p = 2$. This is because both P_t and P_t^* are isometries for all $t \in \mathbb{R}$.

Perron–Frobenius and Koopman operators may also be associated with semidynamical systems $(\alpha_t, t \geq 0)$. In this case we define $P_t: L^1(S, \mu) \to L^1(S, \mu)$ by

$$\int_A P_t f(x)\mu(dx) = \int_{\alpha_t^{-1}(A)} f(x)\mu(dx),$$

for all $t \geq 0$, $A \in \mathcal{B}(S)$, $f \in L^1(S, \mu)$. This clearly agrees with (8.2.4) when we have a quasi-invariant dynamical system. In general this prescription is not enough to deliver a C_0-semigroup; only (S1) and (S2) are satisfied, and so we only have an AO semigroup. For more details, see section 7.4 of Lasota and Mackey [62].

Remarks 1. This process of lifting a semidynamical/dynamical system to a semigroup/group of operators is called "Koopmanism" in recognition of the pioneering work of Bernard O. Koopman.

2. If μ is an invariant measure, then for all $t \in \mathbb{R}$, $f \in C_c(S)$, $x \in S$,

$$P_t f(x) = f(\alpha_{-t}(x)), \quad U_t f(x) = f(\alpha_t(x)).$$

3. Even if μ is not an invariant measure for the dynamical system, it always is for the group, i.e.,

$$\int_S P_t f(x) \mu(dx) = \int_S f(x) \mu(dx),$$

for all $t \in \mathbb{R}$, $f \in B_b^+(S)$. This is proved by essentially the same argument as that used to show that P_t is an L^1-isometry.

4. The Perron–Frobenius operators are positive in that if $f \geq 0$ (a.e.), then for all $t \in \mathbb{R}$, $P_t f \geq 0$ (a.e.).[3]

We now give a swift application of Koopmanism to the ergodic theory of dynamical systems. We say that an invariant probability measure μ is *ergodic* for the semidynamical system $(\alpha_t, t \geq 0)$ if $A \in \mathcal{B}(S)$ with $\alpha_t^{-1}(A) = A$ for all $t \in \mathbb{R}$ implies that $\mu(A) = 0$ or $\mu(A) = 1$. The next theorem is based on Lasota and Mackey [62], pp. 215–6.

Theorem 8.2.8 *An invariant probability measure μ is ergodic for a semidynamical system $(\alpha_t, t \geq 0)$ if and only if the only fixed points of U_t are constant functions, for all $t \geq 0$.*

Proof. First assume that $(\alpha_t, t \geq 0)$ is not ergodic, so there exists a non-trivial set $A \subset S$ such that $\alpha_t^{-1}(A) = A$ for all $t \geq 0$. Let $f := 1_A$, then

$$U_t f = f \circ \alpha_t = 1_{\alpha_t^{-1}(A)} = 1_A = f,$$

and so U_t has a non-constant fixed point. Conversely, assume that U_t has a non-constant fixed point f for some $t > 0$. Then there exists $r > 0$ such that $A := f^{-1}((r, \infty))$ is non-trivial. But then

---

[3] Positive operators that are L^1-isometries are called *Markov operators* in Lasota and Mackey [62].

$$\alpha_t^{-1}(A) = \{x \in S; \alpha_t(x) \in A\}$$
$$= \{x \in S; f(\alpha_t(x)) < r\}$$
$$= \{x \in S; U_t f(x) < r\}$$
$$= \{x \in S; f(x) < r\} = A,$$

so A is invariant, and the semidynamical system cannot be ergodic. $\square$

8.2.2 Dynamical Systems and Differential Equations

Let $b: \mathbb{R}^d \rightarrow \mathbb{R}^d$ and consider the first-order differential equation (ODE) given by

$$\frac{dc(t)}{dt} = b(c(t)), \tag{8.2.7}$$

with initial condition $c(0) = x \in \mathbb{R}^d$. Of course, (8.2.7) is a vector-valued ODE which can be read component-wise as

$$\frac{dc_i(t)}{dt} = b_i(c(t)),$$

for $1 \leq i \leq d$.

If b satisfies a global Lipschitz condition whereby

$$|b(y_1) - b(y_2)| \leq K|y_1 - y_2|, \tag{8.2.8}$$

for all $y_1, y_2 \in \mathbb{R}^d$, where $K > 0$, then there exists a unique solution $(c_x(t), t \in \mathbb{R})$, which is jointly continuous in both variables. Consider the family of mappings $\phi_t: \mathbb{R}^d \rightarrow \mathbb{R}^d$, where $t \in \mathbb{R}$, defined for each $x \in \mathbb{R}^d$ by

$$\phi_t(x) = c_x(t).$$

We call $(\phi_t, t \in \mathbb{R})$ the *(solution) flow* associated to the ODE (8.2.7). It can be helpful to think of ϕ_t as mapping the input x to the output $c_x(t)$ at time t.

Proposition 8.2.9 *The solution flow to (8.2.7) satisfies $\phi_{s+t} = \phi_s \circ \phi_s$ for all $s, t \in \mathbb{R}$.*

Proof. Integrating (8.2.7), we have

$$\phi_{s+t}(x) = x + \int_0^{s+t} b(\phi_u(x))du$$
$$= x + \int_0^s b(\phi_u(x))du + \int_s^{s+t} b(\phi_u(x))du$$
$$= \phi_s(x) + \int_0^t \phi_{u+s}(x)du.$$

But we also have

$$\phi_t(\phi_s(x)) = \phi_s(x) + \int_0^t b(\phi_u(\phi_s(x)))du,$$

and the result then follows by uniqueness of solutions to (8.2.7).   □

It follows that our solution flow is a dynamical system. We say that $(\phi_t, t \in \mathbb{R})$ is a flow of C^k-diffeomorphisms (for $k \in \mathbb{Z}_+$) of $\mathbb{R}^d$ if ϕ_t is a C^k-diffeomorphism[4] for each $t \in \mathbb{R}$. It is shown in, e.g., section 6.1 of Applebaum [6] that if $b \in C_b^k(\mathbb{R}^d)$, then $(\phi_t, t \in \mathbb{R})$ is a flow of C^k-diffeomorphisms. If we drop the condition that b is globally Lipschitz, and impose the weaker local Lipschitz condition that (8.2.8) only holds for x, y in some compact interval of $[a, b]$, then a unique solution to (8.2.7) may only exist for $t \in [-T, T]$ for some $T > 0$. From another viewpoint, we obtain a semidynamical system if a unique solution to (8.2.7) exists only for $t \geq 0$. In this case the mappings $(\phi_t, t \geq 0)$ are called a *semiflow*. For example, the celebrated Lorentz equations, which describe convective heating from below of a fluid, give rise to a semiflow on $\mathbb{R}^3$, as shown in Milani and Koksch [67], pp. 46–8. Here we have $\phi_t(x, y, z) = (x(t), y(t), z(t))$, for all $(x, y, z) \in \mathbb{R}^3, t \geq 0$, where

$$\frac{dx}{dt} = -\sigma x + \sigma y$$

$$\frac{dy}{dt} = rx - y - xz$$

$$\frac{dz}{dt} = -bz + xy,$$

with $\sigma, r, b > 0$.

The ideas we have just discussed about flows extend naturally to the generalisation from $\mathbb{R}^d$ to a d-dimensional smooth manifold M equipped with a Borel measure μ. If X is a C^∞ vector field on M, we may consider the ODE:

$$\frac{dc(t)}{dt} = X(c(t)). \qquad (8.2.9)$$

In a point p belonging to a local co-ordinate patch U of M, we may write $X(p) = \sum_{i=1}^d b^i(p)\partial_i$, and so (8.2.9) reduces to (8.2.7) on U. We say that the vector field X is *complete* if (8.2.9) has a unique solution curve $(c_x(t), t \in \mathbb{R})$. In that case we again obtain a flow of diffeomorphisms $(\phi_t, t \in \mathbb{R})$ (see Abraham et al. [1] for details), where $\phi_t(x) = c_x(t)$, for $x \in M, t \in \mathbb{R}$, and this flow is a dynamical system. From now on, we assume that X is complete. Now consider the real Hilbert space $L^2(M, \mu)$. Assuming that $(\phi_t, t \in \mathbb{R})$ is

---

[4] The case $k = 0$ corresponds to a flow of homeomorphisms.

a continuously quasi-invariant dynamical system, we obtain a one-parameter unitary group of Perron–Frobenius operators $(P_t, t \in \mathbb{R})$ acting in $L^2(M, \mu)$. So for all $t \in \mathbb{R}$, $f \in C_c(M)$, $P_t f = (f \circ \phi_t)(g_t^\mu)^{\frac{1}{2}}$. From now on we will assume that our dynamical system is also differentiably quasi-invariant, and we define the *divergence* of the vector field X with respect to the measure μ by the prescription:

$$\operatorname{div}_\mu(X) := \frac{d}{dt} g_t^\mu(\cdot) \Big|_{t=0}.$$

Then $\operatorname{div}_\mu(X)$ is a continuous function on M. We will see below how this notion of divergence agrees with standard notions from differential geometry, where natural examples of this set-up occur. We define a linear operator T_X in $L^2(M, \mu)$ with domain $C_c^\infty(M)$ by the prescription

$$T_X f = -Xf + \frac{1}{2}\operatorname{div}_\mu(X)f,$$

for all $f \in C_c^\infty(M)$.

Theorem 8.2.10 *The operator T_X is closeable and its closure is the infinitesimal generator of $(P_t, t \in \mathbb{R})$. In particular, T_X is essentially skew-adjoint.*

Proof. Let A be the generator of the unitary group $(P_t, t \in \mathbb{R})$. Then by Stone's theorem (Theorem 4.3.11), A is skew-adjoint, and for all $f, h \in C_c^\infty(M)$, we have

$$\frac{d}{dt} P_t f \Big|_{t=0} = \frac{d}{dt}((f \circ \phi_{-t})(g_t^\mu)^{\frac{1}{2}}) \Big|_{t=0}$$

$$= -Xf + \frac{1}{2}\operatorname{div}_\mu(X)f = T_X f,$$

and so $C_c^\infty(M) \subseteq D_A$ and $Af = T_X f$ for all $f \in C_c^\infty(M)$. By Theorem 1.3.18, $C_c^\infty(M)$ is a core for A and hence $\overline{T_X} = A$. Since $P_t^* = P_{-t}$ for all $t \in \mathbb{R}$, it is an easy calculation to show that T_X is skew-symmetric on $C_c^\infty(M)$. Hence $-\overline{T_X} \subseteq T_X^*$. On the other hand, for all $t \geq 0$, $f \in C_c^\infty(M)$, $g \in D_{T_X^*}$,

$$0 = \langle P_t f - f - \int_0^t T_X P_s f ds, g \rangle$$

$$= \langle f, P_t^* g - g - \int_0^t P_s^* T_X^* g ds \rangle.$$

Hence, by density, $P_t^* g - g - \int_0^t P_s^* T_X^* g = 0$, from which it follows that $g \in D_{-A}$. Then $T_X^* \subseteq -A = -\overline{T_X}$. So we have shown that $T_X^* = -A = -\overline{T_X}$. Hence T_X is essentially skew-adjoint. $\square$

When μ is an invariant measure for $(P_t, t \in \mathbb{R})$, $\mathrm{div}_\mu(X) = 0$, and $T_X f = -Xf$, for all $f \in C_c^\infty(M)$.

A natural context for the above discussion is when M is an orientable manifold and μ is a *volume measure*. Then μ is the unique Borel measure that is naturally associated to a volume form on M (i.e., a positively oriented smooth d-form), which we also call μ. Locally we have $L_X \mu = -\mathrm{div}_\mu(X)\mu$, where L_X is the Lie derivative of X acting on the space of smooth d-forms; globally we define the *Jacobian determinant* $J_{t,\mu} \in C^\infty(M)$ by $\phi_t^* \mu = (J_{t,\mu})\mu$, where $*$ denotes the pullback to the space of smooth d-forms. Then $\mathrm{div}_\mu(X) = -\frac{d}{dt} J_{t,\mu}\big|_{t=0}$ (see pp. 454–5 and p. 471 of Abraham et al. [1] for more details), and the Radon–Nikodym derivatives g_t^μ are Jacobian determinants. This is the promised geometric interpretation of the divergence.

In this context, where μ is a volume form, there is an interesting converse to Theorem 8.2.10. Assume that X is a smooth vector field and consider the linear operator T_X defined on $C_c^\infty(M)$. The *Povzner–Nelson theorem* states that if T_X is essentially skew-adjoint, then the vector field X is complete almost everywhere in that there exists a set C of μ-measure zero, such that the corresponding dynamical system ϕ_t gives rise to a flow on $M \setminus C$ for all $t \in \mathbb{R}$. For details, see section 7.4 of Abraham et al. [1].

An important example is given by Hamiltonian mechanics. We could look at this on a manifold, but for simplicity let us return to a Euclidean set-up. So let $\mathbb{R}^{2d}$ be the phase space for a mechanical system having generalised coordinates $(q_1, \ldots, q_d, p_1, \ldots p_d)$. If $H \colon \mathbb{R}^{2d} \to \mathbb{R}$ is the Hamiltonian (total energy) of the system, then we have Hamilton's equations:

$$\dot{q}_i = \frac{\partial H}{\partial p_i}, \quad \dot{p}_i = -\frac{\partial H}{\partial q_i},$$

for $i = 1, \ldots, d$. Then we obtain a dynamical system by the prescription $\alpha_t((q_1, \ldots, q_d, p_1, \ldots p_d)) = (q_1(t), \ldots, q_d(t), p_1(t), \ldots p_d(t))$, for each $t \in \mathbb{R}$. Assume that $(\alpha_t, t \in \mathbb{R})$ is a differentiably quasi-invariant dynamical system for which the Radon–Nikodym derivatives of the measures μ_t, for $t \in \mathbb{R}$, with respect to Lebesgue measure λ on $\mathbb{R}^{2d}$ are given by $d\mu_t / d\lambda = \rho_t$. Then *Liouville's theorem* tells us that

$$\frac{\partial \rho_t}{\partial t} = \{H, \rho_t\},$$

where the braces define the *Poisson bracket*

$$\{H, \rho_t\} := \sum_{i=1}^d \left(\frac{\partial H}{\partial q_i} \frac{\partial \rho_t}{\partial p_i} - \frac{\partial H}{\partial p_i} \frac{\partial \rho_t}{\partial q_i} \right).$$

Clearly we have that ρ_t is constant in time if and only if $\{H, \rho_t\} = 0$ for all $t \in \mathbb{R}$. It then follows that constant multiples of Lebesgue measure on phase space are invariant measures for the Hamiltonian system. For details of Hamiltonian mechanics and Liouville's equation, see, e.g., Goldstein [39].

Having introduced quantum mechanics earlier, and Poisson brackets above, it is worth pointing out that in his influential book [29], Paul Dirac proposed a quantisation scheme, whereby if $f, g \colon \mathbb{R}^{2d} \to \mathbb{R}$ are classical observables in phase space, and if Q_f, Q_g are the corresponding quantum observables acting in $L^2(\mathbb{R}^d)$, then in going from classical mechanics to quantum mechanics, Poisson brackets should be replaced by commutators. To be precise:

$$\{f, g\} \to -\frac{i}{\hbar}[Q_f, Q_g].$$

Trying to make rigorous sense of this has led to some interesting mathematics; see, e.g., Woodhouse [100].

8.3 Approaches to Irreversibility

There are at least two beliefs held by physicists concerning the essence of irreversible phenomena. The first, and most widely held, is that these are not fundamental. The "true laws" or "theory of everything" of physics will be of the same nature as Newton's laws or Schrödinger's equation in that they will be invariant under time symmetry $t \to -t$. Indeed, classical dynamics treats time inversion as equivalent to velocity inversion in positive time, so that if a particle moves from a to b in time $t > 0$ with initial velocity v, then running the process back in time is the same as moving from b to a in positive time, but with initial velocity $-v$. What we see as irreversible motion within the bulk properties of matter is an averaging out of a vast number of deterministic trajectories of microscopic particles, which human limitations prevent us from being able to observe. The other point of view begins from the human within the world as they find it and takes irreversible motion as a given. From that point of view, reversible dynamics is an idealisation which is not found in the natural world. A very broad and erudite account of both the scientific and philosophical foundations of this latter viewpoint can be found in Prigogine and Stengers [75]. In the next two subsections, we will look at schemes that give us ways of thinking mathematically about these two contrasting approaches.

8.3.1 Dilations of Semigroups

Here we are given a semigroup, and we will seek to construct a group on a larger space from which the semigroup can be obtained by averaging out the "ambient noise". We take a naive approach in order to illustrate the idea in the simplest possible context. We make no claim that this particular scheme is of any physical relevance. To be precise, let $(T_t, t \geq 0)$ be a self-adjoint semigroup acting in a Hilbert space H, with generator $-A$, where A is a positive self-adjoint operator in H (not necessarily bounded). We think of the semigroup as describing our irreversible phenomenon. We could extend it to a group, as in Chapter 4, but we want to go further than this and understand the semigroup as arising from the averaging of a group action which acts in a larger space so that we can incorporate the "noise" of microscopic interactions. We do this in an ad hoc way by introducing a Brownian motion process $(B(t), t \geq 0)$ on a probability (path) space $(\Omega, \mathcal{F}, P)$, which we realise in a canonical manner as $B(t)\omega = \omega(t)$, for all $t \geq 0, \omega \in \Omega$ (see Appendix D). Our starting point is the simple identity, for $s, t \geq 0$

$$B(s + t) = B(s) + (B(s + t) - B(s))$$
$$= B(s) + \theta(s)B(t), \tag{8.3.10}$$

where $(\theta(s), s \geq 0)$ is the (measure-preserving) group of shifts in Ω, given by

$$\theta(s)\omega(t) = \omega(s + t) - \omega(s).$$

Then we induce a unitary group action on $L^2(\Omega, \mathcal{F}, P)$ by the action

$$\Gamma(s)F = F \circ \theta(s),$$

for $F \in L^2(\Omega, \mathcal{F}, P)$ (see, e.g., Appendix 2.10, pp. 141–2 in Applebaum [6] for a proof of unitarity). Let $A^{\frac{1}{2}}$ be the positive self-adjoint square root of A. Note that $D_A \subseteq D_{A^{\frac{1}{2}}}$, since for all $\psi \in D_A$,

$$||A^{\frac{1}{2}}\psi||^2 = \langle A\psi, \psi \rangle < \infty.$$

If $(P(\lambda), \lambda \in \mathbb{R})$ is the resolution of the identity associated to A through the spectral theorem, we have

$$A^{\frac{1}{2}}\psi = \int_{\mathbb{R}} \sqrt{\lambda} P(d\lambda)\psi,$$

for all $\psi \in D_A$.

Next we consider the family of unitary operators $(U_B(t), t \geq 0)$ on $L^2(\Omega, H)$ from (4.4.9), given for all $t \in \mathbb{R}$ by

$$U_B(t) = e^{iB(t)A^{\frac{1}{2}}} = \int_{\mathbb{R}} e^{i\lambda^{\frac{1}{2}}B(t)} P(d\lambda).$$

This is not a semigroup, but we do have the following *cocycle identity* for all $s, t \geq 0$, $F \in L^2(\Omega, H)$,

$$\begin{aligned} U(s+t)F &= e^{iB(s+t)A^{\frac{1}{2}}} F \\ &= e^{iB(s)A^{\frac{1}{2}}} e^{i\Gamma(s)B(t)\Gamma(s)^{-1}A^{\frac{1}{2}}} F \\ &= e^{iB(s)A^{\frac{1}{2}}} \Gamma(s) e^{iB(t)A^{\frac{1}{2}}} \Gamma(s)^{-1} F \\ &= U(s)\Gamma(s)U(t)\Gamma(s)^{-1}, \end{aligned} \qquad (8.3.11)$$

where we have used (8.3.10).

Next, for each $t \geq 0$, we define $W(t) = U(t)\Gamma(t)$. Then $(W(t), t \geq 0)$ is a C_0-semigroup comprising unitary operators. We will not prove strong continuity here, but observe that the semigroup property (S1) follows directly from (8.3.11) as for all $s, t \geq 0$,

$$W(s+t) = U(s+t)\Gamma(s+t) = U(s)\Gamma(s)U(t)\Gamma(s)^{-1}\Gamma(s)\Gamma(t) = W(s)W(t).$$

Then $(W(t), t \geq 0)$ extends to a unitary group, which we write as $(W(t), t \geq 0)$ by the prescription (4.3.3). Next we introduce the natural embedding ι of H into $L^2(\Omega, H)$, which maps each $\psi \in H$ to the path which takes constant value ψ.

Now by Fubini's theorem, and using the characteristic function of Brownian motion from section 4.1, we have for all $\phi, \psi \in H, t \geq 0$,

$$\begin{aligned} \mathbb{E}(\langle W(t)\iota(\phi), \iota\psi \rangle) &= \mathbb{E}(\langle U(t)\iota(\phi), \iota(\psi) \rangle) \\ &= \int_{\mathbb{R}} \mathbb{E}(e^{i\lambda^{\frac{1}{2}}B(t)})\langle P(d\lambda)\phi, \psi \rangle \\ &= \int_{\mathbb{R}} e^{-t\lambda}\langle P(d\lambda)\phi, \psi \rangle \\ &= \langle T_t\phi, \psi \rangle, \end{aligned}$$

from which we deduce that

$$\mathbb{E}(W(t) \circ \iota) = T_t,$$

and so we have *dilated* the semigroup to a group action, so that the semigroup is obtained by averaging out over all the trajectories of Brownian motion; in physics, this procedure is called *coarse graining*. This can be summed up in the following commuting diagram which holds for all $t \geq 0$:

$$
\begin{array}{ccc}
H & \xrightarrow{\ T_t\ } & H \\
\iota \downarrow & & \mathbb{E} \uparrow \\
L^2(\Omega, H) & \xrightarrow{\ W(t)\ } & L^2(\Omega, H).
\end{array}
$$

We have given a very simple example of dilations here. The methodology underlines the way in which stochastic differential equations average out to yield Feller semigroups (see, e.g., Chapter 6 of Applebaum [6]) and the quantum dynamical semigroups introduced at the end of Chapter 5 may also be obtained from solutions of quantum stochastic differential equations averaged out using the vacuum conditional expectation (see, e.g., Parthasarathy [73]) or by using classical Brownian motion, as in Alicki and Fannes [5].

The term *dilation* comes from the foundational work of Béla Sz-Nagy who proved that if T is a contraction in a Hilbert space, then there exists a unitary operator U in a larger Hilbert space H', such that $T^n P = P U^n$ for all $n \in \mathbb{N}$, where P is the projection from H' to H (see Foias and Sz-Nagy [37]). In Chapter 6 of Davies [25], this is extended to contraction semigroups $(T_t, t \geq 0)$ in H, i.e., it is shown that there exists a one-parameter unitary group $(U_t, t \in \mathbb{R})$ on H', so that $T_t P = P U_t$ for all $t \geq 0$.

8.3.2 The Misra–Prigogine–Courbage Approach

In this small subsection, we will do little more than bring the reader's attention to an alternative approach to coarse graining for relating irreversible and reversible dynamics. This was developed by Ilya Prigogine and colleagues in Misra et al. [68], and has since been worked on by many authors, such as Suchanecki [96]. Let us suppose that $(\alpha_t, t \in \mathbb{R})$ is a dynamical system on a locally compact space S equipped with an invariant measure μ on a given σ-algebra $\mathcal{A}$. We may then associate to our system the group of Koopman operators $(U_t, t \in \mathbb{R})$ acting on $L^2(S, \mu)$ that is given by (8.2.6). Now Misra et al. in [68] are interested in systems for which there is a nonunitary bounded invertible linear operator Λ in $L^2(S, \mu)$ such that there is a Markov semigroup $(T_t, t \geq 0)$ in $L^2(S, \mu)$ for which

$$
T_t^* = \Lambda U_t \Lambda^{-1}, \tag{8.3.12}
$$

for all $t \geq 0$. A rich class of systems, called *K-systems*, for which this works comes from ergodic theory. These are precisely those systems for which there exists a sub-σ-algebra $\mathcal{A}_0 \subseteq \mathcal{A}$, for which, writing $\mathcal{A}_t := \alpha_t(\mathcal{A}_0)$ for all $t \in \mathbb{R}$,

1. $\mathcal{A}_s \subseteq \mathcal{A}_t$ for all $-\infty < s \leq t < \infty$.
2. $\bigcup_{t=-\infty}^{\infty} \mathcal{A}_t = \mathcal{A}$.
3. $\bigcap_{t=-\infty}^{\infty} \mathcal{A}_t$ is trivial.

These form a rich class of ergodic dynamical systems which, in particular, contain all Bernoulli shifts. The key point from our perspective is that Λ is seen as a change of representation rather than an averaging process, so no information is lost in moving from the dynamical systems perspective to the Markov semigroup point of view.

9

Varopoulos Semigroups

9.1 Prelude – Fractional Calculus

We set the scene for this chapter by introducing the basics of fractional calculus. The reason for this is that we will be interested in generalisations of fractional integral operators, and we will see that these are more closely related to semigroup theory than may appear to be the case at first glance.

Let us begin with some naive motivation for fractional differentiation. In this section, D will mean the operator of differentiation with respect to a single real variable x, so we are no longer using our usual PDE notation. We start with the simple fact that for all natural numbers $n > m$ and positive real numbers x,

$$D^m x^n = \frac{n!}{(n-m)!} x^{n-m}.$$

Then, if $\beta > \alpha > 0$ are rational numbers, an obvious generalisation is to define the fractional derivative D^α to have the property:

$$D^\alpha x^\beta = \frac{\Gamma(\beta+1)}{\Gamma(\beta+1+\alpha)} x^{\beta-\alpha}. \tag{9.1.1}$$

Now let us turn to integration and define the nth iterated integral of a suitable (locally integrable) function $f : \mathbb{R} \to \mathbb{R}$ at $x \in \mathbb{R}$ by

$$I_n f(x) = \int_0^x \int_0^{y_1} \cdots \int_0^{y_{n-1}} f(y_n) dy_n dy_{n-1} \ldots dy_1.$$

Cauchy showed that

$$I_n f(x) = \frac{1}{(n-1)!} \int_0^x f(t)(x-t)^{n-1} dt,$$

and this is easily verified by induction. Riemann generalised this to fractional order $\alpha > 0$ and defined what is now called the *Riemann–Liouville integral*,

$$I_\alpha f(x) = \frac{1}{\Gamma(\alpha)} \int_0^x f(t)(x-t)^{\alpha-1} dt. \qquad (9.1.2)$$

It is an easy exercise to check that the semigroup property $I_{\alpha+\beta} = I_\alpha I_\beta$ holds, and so for arbitrary positive rational numbers r, we have $I_r = I_{[r]} I_{\{r\}}$, where $[r]$ is the integer part of r and $\{r\}$ is its fractional part. So we really only need to use (9.1.2) when $0 < \alpha < 1$. It is a straightforward exercise to compute

$$I_\alpha x^\beta = \frac{\Gamma(\beta+1)}{\Gamma(\beta+1+\alpha)} x^{\beta+\alpha},$$

and it is interesting to compare this with (9.1.1). In fact, it is natural to seek to define the fractional derivative D^α to be the left inverse of I_α so that $D^\alpha I_\alpha f = f$ for all locally integrable f. However, since for all $m \in \mathbb{N}, D^m I_m f = f$, it is more useful to define

$$D^\alpha = D^m I_{m-\alpha}, \qquad (9.1.3)$$

whenever $m - 1 < \alpha \le m$.

We call D^α the *Riemann–Liouville fractional derivative*. Combining (9.1.2) and (9.1.3), we get for all locally integrable f:

$$(D^\alpha f)(x) = \frac{1}{\Gamma(m-\alpha)} \frac{d^m}{dx^m} \int_0^x f(t)(x-t)^{m-\alpha-1} dt, \qquad (9.1.4)$$

and it is then easy to verify that (9.1.1) does indeed hold. A consequence of this is the formula $D^\alpha 1 = \frac{x^{-\alpha}}{\Gamma(1-\alpha)}$ which is not convenient for applied work. To overcome this obstacle, we may introduce a slightly different notion of fractional differentiation, called the *Caputo derivative*,

$$(D_*^\alpha f)(x) = \frac{1}{\Gamma(m-\alpha)} \int_0^x f^m(t)(x-t)^{m-\alpha-1} dt, \qquad (9.1.5)$$

wherein the derivative has been moved inside the integral. This turns out to be much more useful for modelling purposes (see, e.g., Meerschart and Sikowski [66] and references therein). We will not take our investigations of fractional derivatives and integrals beyond this very brief discussion. For more on the historical roots of the subject, see section 1.1 of Oldham and Spanier [72]. Let us proceed to our main business.

9.2 The Hardy–Littlewood–Sobolev Inequality and Riesz Potential Operators

Hardy and Littlewood considered the Riemann–Liouville integral (9.1.2) and asked the question – if $f \in L^p(\mathbb{R})$ for some $p > 1$, what can be said about $I_\alpha f$? The answer they found was that I_α is a bounded linear operator from $L^p(\mathbb{R})$ to $L^q(\mathbb{R})$ where $1/q = 1/p - \alpha$. Sobolev then extended this result to $\mathbb{R}^d$, but in that setting

$$\frac{1}{q} = \frac{1}{p} - \frac{\alpha}{d}. \tag{9.2.6}$$

So the results of Hardy, Littlewood and Sobolev can be summarised as follows: if $p > 1$ and q is given by (9.2.6), then there exists $C > 0$ so that for all $f \in L^p(\mathbb{R}^d)$,

$$\|I_\alpha f\|_q \leq C\|f\|_p. \tag{9.2.7}$$

The inequality (9.2.7) is called the *Hardy–Littlewood–Sobolev inequality*. For the work that follows, we will find it useful to consider a modified version of the d-dimensional generalisation of (9.1.2); however, we will continue to use the notation I_α and trust that no confusion will arise. So from now on we define, for $f \in L^2_{\text{loc}}(\mathbb{R}^d)$,

$$I_\alpha f = \frac{1}{\Gamma(\alpha/2)} \int_{\mathbb{R}^d} f(y)|x - y|^{\alpha - d} dy. \tag{9.2.8}$$

This linear operator is usually called a *Riesz potential operator*. Now recall the heat semigroup acting on $L^p(\mathbb{R}^d)$, and given for $f \in L^p(\mathbb{R}^d)$, $x \in \mathbb{R}^d$, $t \geq 0$ by

$$T_t f(x) = \frac{1}{(2\pi t)^{d/2}} \int_{\mathbb{R}^d} f(y) e^{-|x-y|^2/2t} dy.$$

Proposition 9.2.1 *For all $f \in C_c(\mathbb{R}^d)$ and $x \in \mathbb{R}^d$,*

$$I_\alpha f(x) = C_{d,\alpha} \int_0^\infty t^{\alpha/2-1}(T_t f)(x) dt, \tag{9.2.9}$$

where $C_{d,\alpha} := \dfrac{\Gamma((d - \alpha)/2)}{\pi^{d/2} 2^{\alpha/2} \Gamma(\alpha/2)}$.

Proof. We have $\int_0^\infty t^{\alpha/2-1}(T_t f)(x) dt = \frac{1}{(2\pi)^{d/2}} \int_0^\infty t^{(\alpha-d)/2-1} e^{-|x-y|^2/2t}$ $f(y) dy$. The result follows by use of Fubini's theorem, and the easily established identity

$$\int_0^\infty t^{(\alpha-d)/2-1}e^{-|x-y|^2/2t}dt = 2^{(d-\alpha)/2}\Gamma((d-\alpha)/2)|x-y|^{\alpha-d},$$

which is obtained by a standard change of variable. $\qquad\qquad\square$

It is interesting to observe that the integral on the right-hand side of (9.2.9) is the *Mellin transform* of the semigroup action. Varopoulos [98] had the idea to use (9.2.9) to extend the definition of the Riesz potential operator to a more general class of semigroups. The key ideas that we need for this are introduced in the next section.

9.3 Varopoulos Semigroups

Most of the material that is contained in this section, and the next, is based on Applebaum and Bañuelos [9].

Let $(S, \mathcal{S}, \mu)$ be a measure space and let $L^p(S) := L^p(S, \mathcal{S}, \mu; \mathbb{R})$. We assume that there is a family of linear operators $(T_t, t \geq 0)$ which are contractions and satisfy (S1) and (S2) on $L^p(S)$ for all $1 \leq p \leq \infty$. We further assume that T_t is a self-adjoint C_0-semigroup on $L^2(S)$ for all $t \geq 0$. These are the "standard assumptions" for the remainder of this chapter. They are satisfied by, e.g., self-adjoint convolution semigroups, as discussed in section 3.1.

We say that $(T_t, t \geq 0)$ is a *Varopoulos semigroup*[1] if it satisfies the standard assumptions, and there exists $n > 0$ (which is not required to be an integer) such that for all $1 \leq p < \infty$, there exists $C_{p,n} > 0$ so that for all $t > 0$, $f \in L^p(S)$,

$$||T_t f||_\infty \leq C_{p,n} t^{-\frac{n}{2p}} ||f||_p. \qquad (9.3.10)$$

Following Varopoulos's terminology, the number n will be referred to as the *dimension* of the semigroup T_t.

We next show that to be a Varopoulos semigroup, it is sufficient for (9.3.10) to hold in the case $p = 1$. To prove this we will need the celebrated *Riesz–Thorin interpolation theorem*. We are content with stating this and direct readers to, e.g., Davies [27] for a proof.

Proposition 9.3.2 (The Riesz–Thorin interpolation theorem) *Let* (S, Σ, μ) *be a measure space and* $1 \leq p_0, p_1, q_0, q_1 \leq \infty$ *and* $0 < \lambda < 1$. *Define* p *and* q *by*

$$\frac{1}{p} := \frac{1-\lambda}{p_0} + \frac{\lambda}{p_1} \text{ and } \frac{1}{q} := \frac{1-\lambda}{q_0} + \frac{\lambda}{q_1}.$$

---

[1] These were called (n, p)-ultracontractive semigroups in [9].

Suppose that X is a linear mapping from $L^{p_0}(X) \cap L^{p_1}(X)$ to $L^{q_0}(X) \cap L^{q_1}(X)$ such that there exists $K_0, K_1 \geq 0$ with

$$\|Xf\|_{q_0} \leq K_0 \|f\|_{p_0} \text{ and } \|Xf\|_{q_1} \leq K_0 \|f\|_{p_1},$$

for all $f \in L^{p_0}(X) \cap L^{p_1}(X)$. Then X extends to a bounded linear operator from $L^p(X)$ to $L^q(X)$ with norm at most $K_0^{1-\lambda} K_1^{\lambda}$.

Now suppose that (9.3.10) holds in the case $p = 1$ so that for all $t > 0$,

$$\|T_t f\|_{\infty} \leq C t^{-d/2} \|f\|_1.$$

We extend to $p > 1$ by using Riesz–Thorin interpolation. In fact, using the fact that T_t is a contraction on $L^{\infty}(\mathbb{R}^d)$, in Proposition 9.3.2 we take $p_0 = q_0 = q_1 = \infty$ and $p_1 = 1$. Choosing $\lambda = 1/p$, we then obtain

$$\|T_t f\|_{\infty} \leq C t^{-d/2p} \|f\|_p,$$

as required.

We next introduce a wider class of semigroups. Suppose that $(T_t, t \geq 0)$ satisfies the standard assumptions. We say that it is *ultracontractive* if $T_t : L^2(S) \to L^{\infty}(S)$ for all $t > 0$. Clearly all Varopoulos semigroups are ultracontractive. We state some useful results on positive (i.e., that for all $f \in L^2(S)$ with $f \geq 0$ (a.e.), we have $T_t f \geq 0$ (a.e.) for all $t > 0$) ultracontractive semigroups, which are proved in Davies [26], pp. 59–61.

- Every positive ultracontractive semigroup $(T_t, t \geq 0)$ has a symmetric kernel $k : (0, \infty) \times S \times S \to [0, \infty)$ so that

$$T_t f(x) = \int_S f(y) k_t(x, y) \mu(dy),$$

for all $f \in L^p(S)$, $x \in S, t > 0$ and moreover

$$\sup_{x, y \in S} k_t(x, y) \leq c_t,$$

where the mapping $t \to c_t$ is monotonic decreasing on $(0, \infty)$ with $\lim_{t \to 0} c_t = \infty$.

- If $\mu(S) < \infty$, then $T_t : L^p(S) \to L^p(S)$ is compact for all $t > 0, 1 \leq p \leq \infty$.

We next explain how to establish a partial converse to the first of these properties. Suppose we have a semigroup $(T_t, t \geq 0)$ that satisfies the standard assumptions and is given by a kernel, so that

$$T_t f(x) = \int_S f(y) k_t(x, y) \mu(dy)$$

for all $x \in S$, $f \in L^p(S)$, $1 \leq p \leq \infty$. What conditions on the kernel will ensure that the semigroup is Varopoulos?

Assume that the kernel $k \in C((0, \infty) \times S \times S)$ and is also such that

- $\int_S k_t(x, y)\mu(dy) = 1$ for all $t > 0$, $x \in S$ (so that $k_t(x, \cdot)$ is the density, with respect to the reference measure μ, of a probability measure on S),
- there exists $C > 0$ so that for all $t > 0$, $x, y \in S$,

$$k_t(x, y) \leq Ct^{-\frac{n}{2}},$$

- k_t is symmetric for all $t > 0$, i.e., $k_t(x, y) = k_t(y, x)$ for all $x, y \in S$.

Then (9.3.10) is satisfied since by Jensen's inequality, for all $1 \leq p < \infty$, $x \in S$, $t > 0$

$$
\begin{aligned}
|T_t f(x)|^p &= \left| \int_S f(y) k_t(x, y)\mu(dy) \right|^p \\
&\leq \int_S |f(y)|^p k_t(x, y)\mu(dy) \\
&\leq Ct^{-\frac{n}{2}} \|f\|_p^p,
\end{aligned}
$$

and so

$$\|T_t f\|_\infty \leq C^{\frac{1}{p}} t^{-\frac{n}{2p}} \|f\|_p.$$

In particular, such conditions on k are satisfied by the heat kernel on certain Riemannian manifolds where $n = d$, the dimension, and on some classes of fractals where $n = 2\frac{\alpha}{\beta}$ where α is the Hausdorff dimension and β is a parameter called the *walk dimension* (see, e.g., Grigor'yan and Telcs [42] for more about these). As shown in Applebaum and Bañuelos [9], it holds for the β-stable transition kernel on Euclidean space, and also on a class of Ricmannian manifolds where $n = \frac{d}{\beta}$. It also holds for uniformly elliptic second-order operators on bounded domains in Euclidean space (see Davies [26], Theorem 2.3.6, pp. 73–4.). We explore this last class of operators, in relation to Varopoulos semigroups, in the last section of this chapter.

In order to generalise the notion of Riesz potential, we will need an additional fact that we will not prove here. This is *Stein's maximal ergodicity lemma* and it states that, under the "standard assumptions", for all $p > 1$, there exists $D_p > 0$ so that for all $f \in L^p(S)$,

$$\|f^*\|_p \leq D_p \|f\|_p, \tag{9.3.11}$$

where for all $x \in S$, $f^*(x) = \sup_{t>0} |T_t f(x)|$. Note also that f^* is a well-defined measurable function. For a proof, we refer readers to the original paper of Stein [93].

From now on we assume that $(T_t, t \geq 0)$ is a Varopoulos semigroup, and note that since the "standard assumptions" are in place, (9.3.11) is also valid. We define the *generalised Riesz potential*

$$I_\alpha f(x) = \frac{1}{\Gamma(\alpha/2)} \int_0^\infty t^{\alpha/2-1} (T_t f)(x) dt, \qquad (9.3.12)$$

for $f \in L^p(S), x \in S$. So we have effectively replaced the heat semigroup in (9.2.9) by a general Varopoulos semigroup. Our first task is to ensure that the definition makes sense:

Lemma 9.3.3 *The integral defining $I_\alpha(f)$ is absolutely convergent.*

Proof. We split the integral on the right-hand side of (9.3.12) into integrals over the regions $0 \leq t \leq 1$ and $1 < t \leq \infty$. Call these integrals $J_\alpha f(x)$ and $K_\alpha f(x)$, respectively, so that $I_\alpha f(x) = J_\alpha f(x) + K_\alpha f(x)$. Now

$$|J_\alpha f(x)| \leq \frac{1}{\Gamma(\alpha/2)} \int_0^1 t^{\alpha/2-1} f^*(x) dt = \frac{2}{\alpha} \frac{1}{\Gamma(\alpha/2)} f^*(x) < \infty,$$

by finiteness of f^*. Furthermore by (9.3.10) (with $p = 1$),

$$|K_\alpha f(x)| \leq C_{1,n} \frac{\|f\|_1}{\Gamma(\alpha/2)} \int_1^\infty t^{\frac{1}{2}(\alpha-n)-1} dt = \frac{2C_{1,n}\|f\|_1}{(n-\alpha)\Gamma(\alpha/2)} < \infty,$$

and the result follows.   □

To obtain more insight into the role of the operator I_α, let $-A$ be the (self-adjoint) infinitesimal generator of the semigroup $(T_t, t \geq 0)$ and assume that A is a positive operator in $L^2(S)$. For each $\gamma \in \mathbb{R}$, we can construct the self-adjoint operator A^γ in $L^2(S)$ by functional calculus, and we denote its domain in $L^2(S)$ by $\text{Dom}(A^\gamma)$.

Theorem 9.3.4 *Let $\alpha > 0$. For all $f \in \text{Dom}(A^{-\frac{\alpha}{2}}) \cap L^1(S)$,*

$$I_\alpha(f) = A^{-\frac{\alpha}{2}} f,$$

in the sense of linear operators acting on $L^2(S)$.

Proof. We use the spectral theorem to write $T_t = \int_0^\infty e^{-t\lambda} P(d\lambda)$ for all $t \geq 0$ where $P(\cdot)$ is the projection-valued measure associated to A. For all $f \in \text{Dom}(A^{-\frac{\alpha}{2}}), g \in L^2(S)$, we have, using Fubini's theorem,

$$\langle I_\alpha(f), g \rangle = \frac{1}{\Gamma(\alpha/2)} \int_0^\infty \int_0^\infty t^{\alpha/2-1} e^{-\lambda t} \langle P(d\lambda) f, g \rangle dt \qquad (9.3.13)$$

$$= \frac{1}{\Gamma(\alpha/2)} \left(\int_0^\infty t^{\alpha/2-1} e^{-t} dt \right) \left(\int_0^\infty \frac{1}{\lambda^{\frac{\alpha}{2}}} \langle P(d\lambda) f, g \rangle \right)$$

$$= \langle A^{-\frac{\alpha}{2}} f, g \rangle.   □$$

Theorem 9.3.4 tells us that I_α is still a kind of fractional integral, but instead of being the (right) inverse of the fractional derivative D^α, it is the (right) inverse of the fractional power $A^{\alpha/2}$ of the (negation of the) infinitesimal generator of our semigroup.

9.4 Varopoulos's Theorem

We now demonstrate that the Hardy–Littlewood–Sobolev inequality holds for our generalised Riesz potential operators.

In the proof of the following theorem, we will treat $C_{p,n,\alpha}$ as a positive constant, depending only on p, n and α, whose value may vary from line to line. This same methodology will be used without comment in later proofs in this chapter.

Theorem 9.4.5 (Varopoulos' theorem) *Suppose the semigroup T_t has dimension n. Let $0 < \alpha < n$, $1 < p < \frac{n}{\alpha}$ and set $\frac{1}{q} = \frac{1}{p} - \frac{\alpha}{n}$. Then there exists $C_{p,n,\alpha} > 0$ so that for all $f \in L^p(S)$,*

$$\|I_\alpha(f)\|_q \le C_{p,n,\alpha} \|f\|_p. \tag{9.4.14}$$

Proof. Let $\delta > 0$ to be chosen later. Let $x \in S$ be arbitrary and choose $f \in L^1(S) \cap L^p(S)$ with $f \neq 0$. As in the proof of Lemma 9.3.3 we split $I_\alpha f(x) = J_\alpha f(x) + K_\alpha f(x)$ where the integrals on the right-hand side range from 1 to δ and δ to ∞ (respectively). Again arguing as in the proof of Lemma 9.3.3, we find that

$$|J_\alpha f(x)| \le \frac{2}{\alpha} \frac{1}{\Gamma(\alpha/2)} f^*(x) \delta^{\frac{\alpha}{2}}.$$

Now, using (9.3.10) we obtain

$$|K_\alpha f(x)| \le C_{p,n,\alpha} \int_\delta^\infty t^{\frac{\alpha}{2} - \frac{n}{2p} - 1} \|f\|_p$$

$$\le C_{p,n,\alpha} \delta^{\frac{\alpha}{2} - \frac{n}{2p}} \|f\|_p,$$

so that

$$|I_\alpha f(x)| \le C_{p,n,\alpha} (f^*(x) \delta^{\frac{\alpha}{2}} + \delta^{\frac{\alpha}{2} - \frac{n}{2p}} \|f\|_p).$$

Now we use a clever trick that seems to be due to John Nash [70]. Let $G(\delta)$ denote the right-hand side of the last inequality. So far we have specified no value for δ, but now we choose it to be such that $G(\delta)$ is minimised, which is precisely when $G'(\delta) = 0$. An easy calculation yields

$$\delta = \left(\frac{||f||_p}{f^*(x)} \right)^{2p/n},$$

and we then obtain

$$|I_\alpha f(x)| \le C_{p,n,\alpha} \left(f^*(x) \right)^{1 - \alpha p/n} ||f||_p^{\alpha p/n} = C_{p,n,\alpha} \left(f^*(x) \right)^{p/q} ||f||_p^{\alpha p/n}.$$

$$(9.4.15)$$

Thus for $1 < p < \frac{n}{\alpha}$ and using (9.3.11),

$$\begin{aligned}
||I_\alpha f||_q^q &\le C_{p,n,\alpha} ||f||_p^{\alpha pq/n} ||f^*||_p^p \\
&\le C_{p,n,\alpha} ||f||_p^{p(1 + \frac{\alpha q}{n})} \\
&= C_{p,n,\alpha} ||f||_p^q,
\end{aligned}$$

and the required result follows by a density argument.   □

We now show how to obtain a Sobolev-type inequality as a corollary to Theorem 9.4.5.

Corollary 9.4.6 *For all $1 < p < n$, $f \in \mathrm{Dom}(A^{\frac{1}{2}}) \cap L^1(S)$, if $A^{\frac{1}{2}} f \in L^p(S)$, then $f \in L^{\frac{np}{n-p}}(S)$ and*

$$||f||_{\frac{np}{n-p}} \le C_{n,p,1} ||A^{\frac{1}{2}} f||_p.$$

Proof. Take $\alpha = 1$ so that $q = \frac{np}{n-p}$. Applying Theorem 9.3.4 within Theorem 9.4.5 yields $||A^{-\frac{1}{2}} f||_q \le C_{n,p,1} ||f||_p$, and so upon replacing f with $A^{\frac{1}{2}} f$ we find that $||f||_q \le C_{n,p,\alpha} ||A^{\frac{1}{2}} f||_p$ as required.   □

The domain condition in Corollary 9.4.6 may seem somewhat strange, but in most cases of interest the operator A and the space S will be such that $\mathrm{Dom}(A)^{\frac{1}{2}} \cap L^1(S)$ contains a rich set of vectors such as Schwartz space (in $\mathbb{R}^d$) or the smooth functions of compact support (on a manifold) and in practice, we would only apply the inequality to vectors in that set.

Note that in the case where $n > 2$ and $p = 2$ in Corollary 9.4.6 we have

$$||f||_{\frac{2n}{n-2}}^2 \le C_{n,2,1}^2 \mathcal{E}(f),$$

where $\mathcal{E}(f) := \langle Af, f \rangle$ is a Dirichlet form. If S is a complete Riemannian manifold with bounded geometry (that satisfies our assumptions; see below) and $-A$ is the Laplacian Δ, then we have $n = d$, the dimension of the manifold, and the Sobolev inequality of Corollary 9.4.6 takes a more familiar form. See Saloff-Coste [85] for more about this and insight into the important role of Sobolev inequalities within modern analysis.

9.5 Nash Inequality, Symmetric Diffusion Semigroups and Heat Kernels

In this section, we will work in $L^2(\mathbb{R}^d)$ or $L^2(U)$ where $U \subseteq \mathbb{R}^d$ is a bounded domain. We consider the uniformly elliptic diffusion operator

$$
Af(x) = \nabla(a\nabla)f(x)
$$

$$
= \sum_{i,j=1}^{d} \partial_i(a_{ij}(x)\partial_j f(x)),
$$

for $f \in C_c^\infty(\mathbb{R}^d), x \in \mathbb{R}^d$, where the matrix-valued function $a = (a_{ij})$ is uniformly elliptic, symmetric and has smooth entries. We have seen in section 4.1 that its Friedrichs extension generates a positive self-adjoint semigroup $(T_t, t \geq 0)$. Our aim in this section is to show that the semigroup $(T_t, t \geq 0)$ acting on $L^2(U)$ is in the Varopoulos class. We will write $K := \inf\{a(x)v \cdot v; x \in \mathbb{R}^d, ||v|| = 1\}$.

We will follow the excellent notes of Peter Friz [36] from his Part III Cambridge thesis (see also Carlen et al. [21]). First we prove the key Nash inequality.

Theorem 9.5.7 (Nash Inequality) *For all $f \in L^1(\mathbb{R}^d) \cap H^1(\mathbb{R}^d)$, there exists* $c = c(d) > 0$ *so that*

$$
||f||_2^{2+4/d} \leq c||\nabla f||_2^2 ||f||_1^{4/d}. \tag{9.5.16}
$$

Proof. If $f \in C_c^\infty(\mathbb{R}^d)$, we will frequently use the well-known and easily verified facts that for all $y \in \mathbb{R}^d$

$$
\widehat{f}(y) \leq (2\pi)^{-d/2}||f||_1 \text{ and } \int_{\mathbb{R}^d} |\nabla f|^2 dx = \int_{\mathbb{R}^d} |y|^2 |\widehat{f}(y)|^2 dy,
$$

the latter result being a special case of Plancherel's theorem. By Plancherel again, we have

$$
||f||_2^2 = \int_{\mathbb{R}^d} |\widehat{f}(y)|^2 dy
$$

$$
\leq \int_{|y|<R} |\widehat{f}(y)|^2 dy + \int_{|y|\geq R} \frac{|y|^2}{R^2} |\widehat{f}(y)|^2 dy,
$$

for all $R > 0$. From the facts stated above we find that there exists $a > 0$ so that[2]

$$
\int_{|y|<R} \widehat{f}(y)|^2 dy \leq (2\pi)^{-d/2}||f||_1^2 \int_{\mathbb{R}^d} \mathbf{1}_{B_0(R)}(y)dy = a||f||_1^2 R^d,
$$

---

[2] Since the volume of $B_0(R)$ is $\dfrac{\pi^{d/2}}{\Gamma(d/2+1)} R^d$, we in fact have $a = 2^{-d/2}\Gamma(d/2+1)^{-1}$, but we have no need of this explicit value.

and $\int_{|y|\geq R} \frac{|y|^2}{R^2} |\widehat{f}(y)|^2 dy \leq R^{-2}\|\nabla f\|_2^2$, which yields the inequality:

$$\|f\|_2^2 \leq a\|f\|_1^2 R^d + R^{-2}\|\nabla f\|_2^2.$$

Now we play the same trick as we did in the proof of Theorem 9.4.5, although historically it first appeared in this context through the pioneering work of Nash [70]. So we differentiate the right-hand side of the last inequality with respect to R to find the value that minimises it, which is given by

$$R = b\left(\frac{\|\nabla f\|_2}{\|f\|_1}\right)^{2/(d+2)},$$

where $b = (2/ad)^{1/d+2}$. When we substitute back into the last inequality, we obtain (9.5.16) after a straightforward calculation, and the result then follows by a density argument. □

The Nash inequality continues to hold when $\mathbb{R}^d$ is replaced by U. In the case where U is a bounded region and $a = I$, we have a beautiful application of the Nash inequality to show that we have a *spectral gap* for the heat equation on U. To be precise, let $\lambda_0(U)$ denote the infimum of the spectrum of $-\Delta$ on U, i.e.,

$$\lambda_0(U) = \inf_{f\in D_\Delta; f\neq 0} \frac{-\langle \Delta f, f\rangle}{\|f\|_2^2} = \inf_{f\in D_\Delta: f\neq 0} \frac{\|\nabla f\|_2^2}{\|f\|_2^2}.$$

By (9.5.16), we have

$$\frac{\|\nabla f\|_2^2}{\|f\|_2^2} \geq \frac{1}{c} \frac{\|f\|_2^{4/d}}{\|f\|_1^{4/d}}.$$

By the Cauchy–Schwarz inequality, $\|f\|_1 \leq |U|^{1/2}\|f\|_2$, where $|U|$ is the Lebesgue measure of U,[3] and we conclude that

$$\lambda_0(U) \geq \frac{|U|^{-2/d}}{c},$$

and this is a special case of the celebrated *Faber–Krahn inequality*.

Theorem 9.5.8 *The semigroup generated by a uniformly elliptic second-order differential operator, acting in the bounded region U, is in the Varopoulos class.*

Proof. First we choose $f \in C_c^\infty(U)$ with $f \geq 0$ and $\|f\|_1 = 1$. For $t \geq 0$, define $E(t) = \|T_t f\|_2^2$. Then $E(0) = \|f\|_2^2 \geq 0$. The mapping $t \to E(t)$ is

---

[3] The fact that U is bounded is crucial here.

differentiable, and using the uniform ellipticity of the generator, followed by (9.5.16), we find that there exists $C > 0$ so that

$$E(t)' = 2 \int_{\mathbb{R}^d} T_t f(x) \partial_t T_t f(x) dx$$

$$= 2 \int_U T_t f(x) A T_t f(x) dx$$

$$= -2 \int_U \nabla T_t f(x) a(x) \nabla T_t f(x) dx$$

$$\leq -2K \| \nabla T_t f \|_2^2$$

$$\leq -C \| T_t f \|_2^{2+4/d} \| T_t f \|_1^{-4/d}.$$

However, since U is bounded, $1 \in D_A \subseteq L^2(U)$ and $A1 = 0$. Hence $T_t 1 = 1$ for all $t \geq 0$ and so

$$\| T_t f \|_1 = \langle T_t f, 1 \rangle = \langle f, T_t 1 \rangle = \langle f, 1 \rangle = 1.$$

Then the last inequality can be written:

$$E(t)' \leq -C_1 E(t)^{1+2/d},$$

(where $C_1 = Cd/2$), or equivalently,

$$\frac{d E(t)^{-2/d}}{dt} \geq C_1.$$

Integrating we obtain

$$E(t)^{-2/d} - E(0)^{-2/d} \geq C_1 t,$$

and so, since $E(0) \geq 0$, $E(t) \leq C_1 t^{-d/2}$. In other words

$$\| T_t f \|_2 \leq C_1 t^{-d/4} \| f \|_1. \tag{9.5.17}$$

This last inequality extends easily to arbitrary $f \in C_c^\infty(U)$ with $\|f\|_1 = 1$, by writing f as the difference of positive and negative parts. It then extends to arbitrary $f \in L^2(U)$ by a density argument. Now let $g \in L^2(U) \subseteq L^1(U)$ (as U is bounded). Then

$$|\langle T_t f, g \rangle| = |\langle f, T_t g \rangle|$$

$$\leq \| f \|_2 \| T_t g \|_2$$

$$\leq C t^{-d/4} \| f \|_2 \| g \|_1,$$

by (9.5.17). By density of $L^2(U)$ in $L^1(U)$, we have that for all $g \in L^1(U)$,

$$\left| \int_{\mathbb{R}^d} T_t f(x) g(x) dx \right| \leq C t^{-d/4} \| f \|_2 \| g \|_1.$$

Now take the supremum over all $g \in L^1(U)$ such that $||g|| = 1$, and use the fact that $L^1(U)^* = L^\infty(U)$ to find that

$$||T_t f||_\infty \le C_1 t^{-d/4} ||f||_2.$$

But then we have by (9.5.17),

$$||T_t f||_\infty = ||T_{t/2} T_{t/2} f||_\infty$$

$$\le C \left(\frac{t}{2}\right)^{-d/4} ||T_{t/2} f||_2$$

$$\le C \left(\frac{t}{2}\right)^{-d/2} ||f||_1.$$

By density of $L^2(U)$ in $L^1(U)$ this extends to all $f \in L^1(U)$. So we have that for all $t > 0$, $f \in L^1(U)$, there exists $C = C(d) > 0$, so that

$$||T_t f||_\infty \le C t^{-d/2} ||f||_1.$$

It then follows that the semigroup is in the Varopoulos class by the Riesz–Thorin interpolation argument that was presented in section 8.3. $\qquad\square$

Since the semigroup $(T_t, t \ge 0)$ is a Varopoulos one, it is ultracontractive. It is also positive, as can be seen by recognising it as the transition semigroup for the Markov process that is the solution of the stochastic differential equation

$$dY(t) = \sigma(Y(t)) dB(t),$$

which is "killed" on the boundary of U, where $\sigma\sigma^T = a$, and B is a d-dimensional standard Brownian motion. Hence the semigroup has a symmetric "heat kernel" $k_t : U \times U \to [0, \infty)$ for $t > 0$, so that for all $f \in L^2(U), x \in U$,

$$T_t f(x) = \int_U f(y) k_t(x, y) dy.$$

If a is non-constant, we cannot expect there to be a closed formula for k_t, but if we impose Dirichlet boundary conditions, we can obtain *Gaussian bounds* on the kernel. Much of Chapter 3 of Davies [26] is devoted to obtaining these, and thus showing that there exists $C_1, C_2, c_1, c_2 > 0$ so that for all $t > 0, x, y \in U$,

$$C_1 t^{-d/2} \exp\{-|x - y|^2/c_1 t\} \le k_t(x, y) \le C_2 t^{-d/2} \exp\{-|x - y|^2/c_2 t\}.$$

Notes and Further Reading

The founders of the modern analytic theory of semigroups were Einar Hille, William Feller, Isao Miyadera, Ralph Phillips and Kosaku Yosida, and all five of these names are recognised in the first two classification theorems of Chapter 2. Of these five, four are primarily analysts, but Feller is perhaps better known for his ground-breaking contributions to probability, and his work on semigroups (recognised in the term *Feller semigroup*), as we saw in Chapters 3 and 7, indicates the vitality of the link between these areas.

The first published book on the subject was by Hille [45], and this was later expanded into a second edition which was co-authored by Phillips [46]. This was a bible in the subject for many years. Now of course, there is a healthy choice of dedicated books on the subject that give either introductory accounts, such as Engel and Nagel [32], and Sinha and S. Srivasta [91], or more advanced treatments such as Davies [25], Goldstein [40], Pazy [74] or the comprehensive Engel and Nagel [31]. You can also find a lot of material on semigroups in the book by Davies on linear operators [27], wherein Chapters 6, 7 and 8 update key material from the earlier Davies [25]; that by Yosida [101] on functional analysis (see Chapter IX); and that of Kato [59] on perturbation theory (see Chapter 9). If your taste is more in the area of partial differential equations, you can consult Evans [34] (Chapter 7) or Rennardy and Rogers [78] (Chapter 11), and if you are coming from the direction of probability theory and stochastic processes, you might investigate Chapter X of Feller [35], Chapter 4 of Jacob [54] or Chapter 19 of Kallenberg [58]. This list is far from comprehensive, and I apologise to readers for whom I have omitted their favourite source of interesting semigroup-related ideas.

For the remainder of this short conclusion, I will briefly mention some additional important topics in mainstream semigroup theory that were not treated in this volume:

1. *Evolution Systems*

   A natural generalisation of a C_0-semigroup is a *two-parameter evolution system*, i.e., a family of linear operators $(T_{s,t}; 0 \le s \le t < \infty)$ acting in a Banach space E which satisfy the following generalisation of (S2):

   $$T_{s,t} T_{t,u} = T_{s,u},$$

   for all $0 \le s \le t \le u < \infty$, wherein $T_{s,s} = I$ for all $s \ge 0$. Instead of a single operator, the role of the generator is played by a family of linear operators $(A(t), t \ge 0)$ and these are associated to the evolution system through the operator-valued partial differential equations:

   $$\frac{\partial T_{s,t}}{\partial t} = A(t)T_{s,t}; \quad \frac{\partial T_{s,t}}{\partial s} = -A(s)T_{s,t}.$$

   The simplest example is obtained by starting with a family of bounded linear operators $(A(t), t \ge 0)$ for which the mapping $t \to A(t)$ is norm-continuous, and defining for all $0 \le s \le t < \infty$,

   $$T_{s,t} = \exp\left\{ \int_s^t A(r)dr \right\}.$$

   Important examples of where such systems arise include the study of second order parabolic PDEs with time-dependent coefficients, and two-parameter generalisations of semidynamical systems. An excellent source for this topic is Chapter 5 of Pazy [74].

2. *Spectral Theory*

   Spectral theory is a huge subject, and one that we have hardly touched in this book. If $(T_t, t \ge 0)$ is a self-adjoint semigroup acting in a Hilbert space H with bounded (self-adjoint) generator A, the *spectral mapping theorem* tells us that for all $t \ge 0$,

   $$\sigma(T_t) = e^{t\sigma(A)} := \{e^{t\lambda}; \lambda \in \sigma(A)\}.$$

   It is natural to want to extend this to more general classes of C_0-semigroup, and in Chapter 4, section 3 of Engel and Nagel [31] a number of counterexamples are presented to demonstrate that this result does not hold in general. What is valid for all C_0-semigroups is the *spectral inclusion theorem* which states that we always have $\sigma(T_t) \supset e^{t\sigma(A)}$.

3. Asymptotic Behaviour

   Here we are interested in the behaviour of the semigroup as $t \to \infty$. Two natural notions of asymptotic stability are *uniform stability* wherein it is required that $\lim_{t\to\infty} \|T_t\| = 0$, and *uniform exponential stability* for which $\lim_{t\to\infty} e^{\epsilon t} \|T_t\| = 0$ for some $\epsilon > 0$. It can be shown that these notions are equivalent, and a necessary and sufficient condition for either (i.e., both) of them to hold is that $\int_0^\infty ||T_t x||^p dt < \infty$ for all $x \in E$, and for one (equivalently all) $1 \leq p < \infty$. For proofs of these results, and much more about this topic, see Chapter 5 of Engel and Nagel [31].

Asymptotic behaviour is closely related to the study of ergodic averages, and in Chapter 5 of Davies [25], conditions are found for an *ergodic theorem* to hold of the form

$$\lim_{t \to \infty} \frac{1}{t} \int_0^\infty T_s x ds = Px,$$

for all $x \in E$, where $(T_t, t \geq 0)$ is a contraction semigroup in E with generator A, and P is the projection from E onto Ker(A). A sufficient condition for this to hold (for any contraction semigroup therein) is that E is reflexive.

4. Positive Semigroups

A C_0-semigroup $(T_t, t \geq 0)$ in $C_0(\mathbb{R}^d)$ is said to be *positive* if $f \geq 0 \Rightarrow T_t f \geq 0$ for all $t \geq 0$, and we have already seen in Chapter 7 how important this property is in the study of Markov and Feller semigroups. We would like to extend this notion to a more general class of Banach spaces, and the natural context for this is a *Banach lattice E* which is a vector lattice (with order relation $\leq$) that is also a Banach space, and which has the property that for all $x, y \in E$ with $|x| \leq |y|$, we have $||x|| \leq ||y||$ (where $|x| = x \vee -x$). Then we may define positivity for C_0-semigroups $(T_t, t \geq 0)$ in the Banach lattice E, just as we did for $C_0(\mathbb{R}^d)$. It can be shown that $(T_t, t \geq 0)$ is positive if and only if its resolvent $R_\lambda(A)$ is positive for sufficiently large λ. Positivity of semigroups enables extra information to be gained about spectral properties and asymptotic behaviour of semigroups. For more on this topic, see Chapter 5 of Engel and Nagel [32].

Appendix A
The Space $C_0(\mathbb{R}^d)$

In this section we collect together some useful properties of spaces of functions. Let S be a locally compact Hausdorff space. Then we have

$$C_c(S) \subseteq C_0(S) \subseteq C_b(S) \subseteq B_b(S) \subseteq B(S), \qquad (A.0.1)$$

where

- $B(S)$ is the Banach space of all bounded real-valued functions on S, equipped with the norm

$$||f||_\infty := \sup\{|f(x)|, x \in S\}, \text{ for all } f \in B(S).$$

- $B_b(S)$ is the closed subspace of $B(S)$ comprising Borel measurable functions.
- $C_b(S)$ is the closed subspace of $B_b(S)$ comprising bounded continuous functions.
- $C_0(S)$ is the closed subspace of $C_b(S)$ comprising continuous functions that vanish at infinity.
- $C_c(S)$ is the dense subspace of $C_0(S)$ comprising continuous functions with compact support.

Note that both $C_b(S)$ and $C_0(S)$ are Banach spaces in their own right, with respect to the norm $|| \cdot ||_\infty$. If S is a metric space, we may also introduce $UC_b(S)$, which is the closed subspace of $C_b(S)$ comprising bounded uniformly continuous functions, and we have $C_0(S) \subseteq UC_b(S)$.

We prove that $B(S)$ really is a Banach space. To establish this fact, the following result is useful.

Proposition A.0.1 *Let (x_n) be a Cauchy sequence in a metric space (X, d) and let $x \in X$. If (x_n) has a subsequence (x_{n_k}) converging to x, then (x_n) converges to x.*

Proof. Let $\epsilon > 0$. As (x_n) is Cauchy, there exists $N \in \mathbb{N}$ such that $d(x_n, x_m) < \frac{\epsilon}{2}$ for all $n, m > N$. As $x_{n_k} \to x$ as $k \to \infty$, there exists $K \in \mathbb{N}$ such that $d(x_{n_k}, x) < \frac{\epsilon}{2}$ for all $k > K$. Let $M > \max\{K, N\}$ and $m = n_M$. Then $m \geqslant M > N$. Let $n > m$. Then $n, m > N$, so $d(x_n, x_m) < \frac{\epsilon}{2}$. But $x_m = x_{n_M}$ is in the subsequence and $M > K$, so $d(x_m, x) < \frac{\epsilon}{2}$. Then

$$d(x_n, x) \leqslant d(x_n, x_m) + d(x_m, x) < \frac{\epsilon}{2} + \frac{\epsilon}{2} = \epsilon.$$

Thus $x_n \to x$ as $n \to \infty$. $\qquad\square$

Theorem A.0.2 *$B(S)$ is a Banach space and $B_b(S)$ is a closed subspace of $B(S)$.*

Proof. Let (f_n) be a Cauchy sequence in $B(S)$. Then for each $x \in S$, $(f_n(x))$ is a real-valued Cauchy sequence, and so is bounded. Hence by the Bolzano–Weierstrass theorem, it has a convergent subsequence. By Proposition A.0.1, the sequence $(f_n(x))$ also converges to the limit of the subsequence and so we may define

$$f(x) = \lim_{n \to \infty} f_n(x).$$

It is clear that f is a well-defined function on S. We must show that it is the limit of (f_n) in the uniform norm. To see this, first observe that for any $\epsilon > 0$, there exists $N \in \mathbb{N}$ so that $||f_n - f_m||_\infty < \epsilon/2$ for all $m, n > N$. Then for all $x \in S$,

$$\begin{aligned}
|f_n(x) - f(x)| &\leq |f_n(x) - f_m(x)| + |f_m(x) - f(x)| \\
&\leq ||f_n - f_m||_\infty + |f_m(x) - f(x)| \\
&< \epsilon/2 + |f_m(x) - f(x)|.
\end{aligned}$$

We know that $\lim_{m \to \infty} |f_m(x) - f(x)| = 0$. Hence by taking limits as $m \to \infty$ on both sides of the last inequality we deduce that

$$|f_n(x) - f(x)| \leq \epsilon/2 < \epsilon,$$

and so the sequence (f_n) converges uniformly to f, as required.

Finally, f is bounded as, from what has just been proved, given any $\epsilon > 0$, there exists $N \in \mathbb{N}$ so that if $n > N$, then $||f_n - f||_\infty < \epsilon$. But then

$$||f||_\infty \leq ||f_n||_\infty + ||f_n - f||_\infty,$$

and so

$$||f||_\infty \leq \inf_{n > N} ||f_n||_\infty + \epsilon < \infty,$$

and we are done. The result for $B_b(S)$ is an easy exercise in measurability. $\quad\square$

The defining property of $C_0(S)$ is that $f \in C_0(S)$ if and only if for any $\epsilon > 0$ there exists a compact set $K_\epsilon \subseteq S$ so that if $x \in K_\epsilon^c$, then $|f(x)| < \epsilon$. If $S = \mathbb{R}^d$, this is equivalent to the requirement that $\lim_{|x|\to\infty} f(x) = 0$.

Let us now prove some of the inclusions in (A.0.1) that may not be obvious to all readers.

Proposition A.0.3 *If S is a metric space (with metric d), then $C_0(S) \subseteq UC_b(S)$.*

Proof. Let $f \in C_0(S)$. We first show that f is bounded. Take ϵ as in the definition just given. Since a continuous function on a compact set is bounded, we know there exists C_ϵ such that $|f(x)| < C_\epsilon$ for all $x \in K_\epsilon$ and then $|f(x)| \le \max\{C_\epsilon, \epsilon\}$ for all $x \in S$.

For uniform continuity, with ϵ fixed as above, if $x, y \in K_\epsilon$ we can use the fact that continuous functions on compact sets are uniformly continuous. So there exists $\delta > 0$ such that for all $x \in K_\epsilon, y \in B_\delta(x), |f(x) - f(y)| < \epsilon$. If $x, y \in K_\epsilon^c$, then just pick any $\delta > 0$. When $y \in B_\delta(x)$ we have

$$|f(x) - f(y)| \le |f(x)| + |f(y)| < 2\epsilon.$$

Now suppose that $x \in K_\epsilon, y \in K_\epsilon^c$. First note that since f is continuous at any $t \in K_\epsilon$, there exists $0 < \rho_t < 1$ so that if $z \in B_{\rho_t}(t)$, then $|f(t) - f(z)| < \epsilon/2$. The collection of all balls of the form $\{B_{\rho_t/2}(t), t \in K_\epsilon\}$ is an open cover of the compact set K_ϵ, and so we can find a finite subcover comprising balls of radius $\rho_{t_1}/2, \ldots, \rho_{t_N}/2$. Then $x \in B_{\rho_{t_i}/2}(t_i)$ for some $i = 1, \ldots, N$. Let $\delta := \min\{\rho_{t_1}, \ldots, \rho_{t_N}\}$. Then if $y \in B_{\delta/2}(x)$, we have

$$d(y, t_i) \le d(y, x) + d(x, t_i) < \delta/2 + \rho_{t_i}/2 < \rho_{t_i}$$

and so

$$|f(x) - f(y)| \le |f(x) - f(t_i)| + |f(t_i) - f(y)| < \epsilon/2 + \epsilon/2 = \epsilon. \quad \square$$

Proposition A.0.4 *If S is a metric space, then $UC_b(S)$ is closed in $C_b(S)$ and $C_0(S)$ is closed in $UC_b(S)$.*

Proof. For the first statement, let (f_n) be a sequence in $UC_b(S)$ converging to $f \in C_b(S)$. Then given $\epsilon > 0$, there exists $N \in \mathbb{N}$ so that if $n > N, \sup_{x \in S} |f_n(x) - f(x) < \epsilon/3$. Also f_n is uniformly continuous, so for the given ϵ, for all $x \in S$ there exists $\delta_n > 0$ so that if $y \in B_{\delta_n}(x), |f_n(x) - f_n(y)| < \epsilon/3$. Then $f \in UC_b(S)$ since for all $x \in S, y \in B_{\delta_n}(x)$ we have

$$|f(x) - f(y)| \le |f(x) - f_n(x)| + |f_n(x) - f_n(y)| + |f_n(y) - f(y)|$$
$$< \epsilon/3 + \epsilon/3 + \epsilon/3 = \epsilon.$$

So $UC_b(S)$ is closed.

For the second statement, let (g_n) be a sequence in $C_0(S)$ converging to $g \in UC_b(S)$. Then given $\delta > 0$, for each $n \in \mathbb{N}$ there exists a compact set $K_{\delta,n}$ so that $|g(x)| < \epsilon/2$ for $x \in K^c_{\delta,n}$. Also there exists $M \in \mathbb{N}$ so that $\sup_{x \in S} |g_m(x) - g(x)| < \epsilon/2$ for all $m > N$. Hence for such m, and $x \in K^c_{\delta,m}$,

$$|g(x)| \le |g(x) - g_m(x)| + |g_m(x)| < \epsilon/2 + \epsilon/2 = \epsilon,$$

and the result follows. $\qquad\square$

The fact that $C_c(S)$ is dense in $C_0(S)$ is a standard consequence of Uruysohn's lemma in topology. We will not go into this here; a proof of the density result is given in Appendix A of Applebaum [7], p. 181. In the case $S = \mathbb{R}^d$, we can go further and show that $C_c^\infty(\mathbb{R}^d)$ is dense in $C_0(\mathbb{R}^d)$. For a proof of that fact we need the *Stone–Weierstrass theorem*.

Before we state that, we need some facts and definitions. First observe that if S is a locally compact Hausdorff space, then the linear space $C_0(S)$ is in fact an algebra with respect to pointwise multiplication of functions. A subalgebra $\mathcal{A}$ of $C_0(S)$ is said to be *separating* if for all $x, y \in S$ with $x \ne y$, there exists $f \in \mathcal{A}$ such that $f(x) \ne f(y)$ and it *vanishes nowhere* if for each $x \in S$ there exists $f \in \mathcal{A}$ such that $f(x) \ne 0$.

Theorem A.0.5 (Stone–Weierstrass Theorem) *If S is a locally compact Hausdorff space, and $\mathcal{A}$ is a subalgebra of $C_0(S)$ that is separating and vanishes nowhere, then $\mathcal{A}$ is dense in $C_0(S)$.*

For a proof see, e.g., Simmons [89].

To use Theorem A.0.5 to show that $C_c^\infty(\mathbb{R}^d)$ is dense in $C_0(\mathbb{R}^d)$ we introduce a family of functions indexed by $a, b \in \mathbb{R}^d$ with $a \ne b$ defined by

$$\phi_{a,b}(x) = \begin{cases} \exp\left\{ -\dfrac{1}{|a-b|^2 - |2x - a - b|^2} \right\} & \text{if } x \in B_r(p) \\ 0 & \text{if } x \notin B_r(p) \end{cases},$$

where $r := |a - b|/2$ and $p := (a + b)/2$. It is easy to verify that $\phi_{a,b} \in C_c^\infty(\mathbb{R}^d)$ with $\text{supp}(\phi_{a,b}) = \overline{B_r(p)}$. Now $C_c^\infty(\mathbb{R}^d)$ is a subalgebra of $C_0(\mathbb{R}^d)$. To see that it vanishes nowhere, the reader can easily check that for all $x \in \mathbb{R}^d$, there exists $a, b \in \mathbb{R}^d$ with $a \ne b$ such that $\phi_{a,b}(x) > 0$. Finally, to check that it is separating, let $x, y \in \mathbb{R}^d$ with $x \ne y$. Then the reader is invited to choose distinct $a, b \in \mathbb{R}^d$ such that $\phi_{a,b}(x) > 0$ but $|a - b|/2 < |y - x|$ and so $\phi_{a,b}(y) = 0$. It can similarly be shown that if M is a smooth manifold, then $C_c^\infty(M)$ is dense in $C_0(M)$ by using bump functions in place of the $\phi_{a,b}$.

Appendix B

The Fourier Transform

The material in the first part of this section may be found in many texts that include sections on harmonic analysis. This author particularly likes Chapter 7 of Rudin [82], and the proofs given here are based on those found therein.

Let $f \in L^1(\mathbb{R}^d, \mathbb{C})$; then its *Fourier transform* is the mapping $\hat{f}$, defined by

$$\hat{f}(u) = (2\pi)^{-d/2} \int_{\mathbb{R}^d} e^{-i(u,x)} f(x) dx \tag{B.0.1}$$

for all $u \in \mathbb{R}^d$. If we define $\mathcal{F}(f) := \hat{f}$, then $\mathcal{F}$ is a linear mapping from $L^1(\mathbb{R}^d, \mathbb{C})$ to the space of all continuous complex-valued functions on $\mathbb{R}^d$ called the *Fourier transformation*.

We introduce two important families of linear operators in $L^1(\mathbb{R}^d, \mathbb{C})$, *translations* $(\tau_x, x \in \mathbb{R}^d)$ and *phase multiplications* $(e_x, x \in \mathbb{R}^d)$, by

$$(\tau_x f)(y) = f(y - x), \qquad (e_x f)(y) = e^{i(x,y)} f(y)$$

for each $f \in L^1(\mathbb{R}^d, \mathbb{C})$ and $x, y \in \mathbb{R}^d$.

It is easy to show that each of τ_x and e_x are isometric isomorphisms of $L^1(\mathbb{R}^d, \mathbb{C})$. Two key, easily verified properties of the Fourier transform are

$$\widehat{\tau_x f} = e_{-x}\hat{f} \qquad \text{and} \qquad \widehat{e_x f} = \tau_x \hat{f} \tag{B.0.2}$$

for each $x \in \mathbb{R}^d$.

Furthermore, if we define the *convolution* $f * g$ of $f, g \in L^1(\mathbb{R}^d, \mathbb{C})$ by

$$(f * g)(x) = (2\pi)^{-d/2} \int_{\mathbb{R}^d} f(x - y) g(y) dy$$

for each $x \in \mathbb{R}^d$, then we have $\widehat{(f * g)} = \hat{f}\hat{g}$.

If $f \in L^2(\mathbb{R}^d, \mathbb{C})$, we can also define its Fourier transform as in (B.0.1) on the dense linear subspace $C_c(\mathbb{R}^d, \mathbb{C})$ and then by continuous extension to the whole of $L^2(\mathbb{R}^d, \mathbb{C})$. It then transpires that $\mathcal{F}: L^2(\mathbb{R}^d, \mathbb{C}) \to L^2(\mathbb{R}^d, \mathbb{C})$ is a unitary operator. The fact that $\mathcal{F}$ is isometric is sometimes expressed by

Theorem B.0.1 (Plancherel) *If $f \in L^2(\mathbb{R}^d, \mathbb{C})$, then*

$$\int_{\mathbb{R}^d} |f(x)|^2 dx = \int_{\mathbb{R}^d} |\hat{f}(u)|^2 du;$$

or

Theorem B.0.2 (Parseval) *If $f, g \in L^2(\mathbb{R}^d, \mathbb{C})$, then*

$$\int_{\mathbb{R}^d} \overline{f(x)} g(x) dx = \int_{\mathbb{R}^d} \overline{\hat{f}(u)} \hat{g}(u) du.$$

We prove the Parseval and Plancherel theorems below.

Although, as we have seen, $\mathcal{F}$ has nice properties in both L^1 and L^2, perhaps the most natural context in which to discuss it is the Schwartz space $S(\mathbb{R}^d, \mathbb{C})$ of rapidly decreasing functions, as defined in Chapter 3. These are smooth functions such that they, and all their derivatives, decay to zero at infinity faster than any negative power of $|x|$. Note that $S(\mathbb{R}^d, \mathbb{C})$ is dense in $L^p(\mathbb{R}^d, \mathbb{C})$ for all $1 \leq p < \infty$.

The space $S(\mathbb{R}^d, \mathbb{C})$ is a Fréchet space with respect to the family of norms $\{||.||_N, N \in \mathbb{N} \cup \{0\}\}$, where for each $f \in S(\mathbb{R}^d, \mathbb{C})$

$$||f||_N = \max_{|\alpha| \leq N} \sup_{x \in \mathbb{R}^d} (1 + |x|^2)^N |D^\alpha f(x)|.$$

The dual of $S(\mathbb{R}^d, \mathbb{C})$ with this topology is the space $S'(\mathbb{R}^d, \mathbb{C})$ of *tempered distributions*.

If α is a multi-index, we can easily check the standard property of the Fourier transform that for $f \in S(\mathbb{R}^d, \mathbb{C})$, $y \subset \mathbb{R}^d$,

$$\widehat{D^\alpha f}(y) = y^\alpha \hat{f}(y).$$

The operator $\mathcal{F}$ is a continuous bijection of $S(\mathbb{R}^d, \mathbb{C})$ into itself with a continuous inverse. We will not prove all of this here, but we will prove

Lemma B.0.3 *If $f \in S(\mathbb{R}^d, \mathbb{C})$ then $\hat{f} \in S(\mathbb{R}^d, \mathbb{C})$.*

Proof. To show $\hat{f} \in S(\mathbb{R}^d, \mathbb{C})$, we will use the easily established fact that if $f \in L^1(\mathbb{R}^d)$ should be $L^1(\mathbb{R}^d, \mathbb{C})$, then $\hat{f}$ is bounded. If α and β are multi-indices, we first define $g(x) = (-1)^\alpha x^\alpha f(x)$ for all $x \in \mathbb{R}^d$. Then $g \in S(\mathbb{R}^d, \mathbb{C})$ and $\hat{g} = D^\alpha \hat{f}$. Then writing $P^\beta(x) = x^\beta$, with an obvious notation, we have

$$P^\beta D^\alpha \hat{f} = P^\beta \hat{g} = \widehat{D^\beta g},$$

which is bounded, i.e.,

$$\sup_{y\in\mathbb{R}^d} |y^\beta D^\alpha \widehat{f}(y)| < \infty,$$

and we are done.   □

Theorem B.0.4 (Fourier inversion) *If* $f \in S(\mathbb{R}^d, \mathbb{C})$, *then*

$$f(x) = (2\pi)^{-d/2} \int_{\mathbb{R}^d} \widehat{f}(u) e^{i(u,x)} du.$$

Proof. Let $f, g \in L^1(\mathbb{R}^d)$, then by Fubini's theorem,

$$\begin{aligned}
\int_{\mathbb{R}^d} f(x)\widehat{g}(x)dx &= \frac{1}{(2\pi)^{d/2}} \int_{\mathbb{R}^d} f(x) \left(\int_{\mathbb{R}^d} g(y) e^{-ix\cdot y} dy \right) dx \\
&= \frac{1}{(2\pi)^{d/2}} \int_{\mathbb{R}^d} g(y) \left(\int_{\mathbb{R}^d} f(x) e^{-ix\cdot y} dx \right) dy \\
&= \int_{\mathbb{R}^d} g(x)\widehat{f}(x)dx.
\end{aligned}$$

Now take $f, h \in S(\mathbb{R}^d, \mathbb{C})$, and for $\lambda > 0$, let $g(x) = h(x/\lambda)$ in the last identity to find that, by straightforward changes of variable,

$$\begin{aligned}
\int_{\mathbb{R}^d} h(x/\lambda)\widehat{f}(x)dx &= \lambda^d \int_{\mathbb{R}^d} \widehat{h}(\lambda x) f(x)dx \\
&= \int_{\mathbb{R}^d} f(x/\lambda)\widehat{h}(x)dx.
\end{aligned}$$

Now let $\lambda \to \infty$ in the last identity, and use dominated convergence, to find that

$$h(0) \int_{\mathbb{R}^d} \widehat{f}(x)dx = f(0) \int_{\mathbb{R}^d} \widehat{h}(x)dx.$$

Next we take h to be the Gaussian function $h(x) = e^{-|x|^2/2}$, for all $x \in \mathbb{R}^d$. As pointed out above, $h \in S(\mathbb{R}^d, \mathbb{R})$, and we know that $\widehat{h}(x) = h(x)$ for all $x \in \mathbb{R}^d$. Hence we obtain

$$f(0) = \frac{1}{(2\pi)^{d/2}} \int_{\mathbb{R}^d} \widehat{f}(x)dx.$$

The result is then obtained by replacing f with $\tau_{-y} f$ for $y \in \mathbb{R}^d$.   □

Proof of Parveval and Plancherel theorems.

Proof. If $f, g \in S(\mathbb{R}^d, \mathbb{C})$, by Fourier inversion and Fubini's theorem, we have

$$\int_{\mathbb{R}^d} f(x)\overline{g(x)}dx = \frac{1}{(2\pi)^{d/2}} \int_{\mathbb{R}^d} \overline{g(x)} \left(\int_{\mathbb{R}^d} e^{ix\cdot y}\widehat{f}(y)dy \right) dx$$

$$= \int_{\mathbb{R}^d} \widehat{f}(y) \overline{\left(\frac{1}{(2\pi)^{d/2}} \int_{\mathbb{R}^d} e^{-ix\cdot y}g(x)dx \right)} dy$$

$$= \int_{\mathbb{R}^d} \widehat{f}(y)\overline{\widehat{g}(y)}dy.$$

This shows that the operator $\mathcal{F}$ is an isometry from $S(\mathbb{R}^d, \mathbb{C})$ to $S(\mathbb{R}^d, \mathbb{C})$ and so it extends by density to an isometry from $L^2(\mathbb{R}^d, \mathbb{C})$ to itself. Both the Parseval and Plancherel theorems then follow. $\qquad\qquad\square$

Appendix C
Sobolev Spaces

An authoritative reference for Sobolev spaces is the dedicated volume Adams [2]. For our somewhat limited needs, we have followed the succinct account in Rosenberg [81]. Our main interest in these spaces is the role they play as maximal domains of differential operators acting in L^p-spaces. But we will in fact only need the case $p = 2$. Let U be a bounded open subset of $\mathbb{R}^d$. We define a norm on the space $C^m(U)$ by the prescription

$$\|f\|_{H^m(U)} := \left(\int_U \sum_{|\alpha| \le m} D^\alpha f(x)^2 dx \right)^{\frac{1}{2}}.$$

The completion of $C^m(U)$ under this norm is called the *Sobolev space of order m*. It is a (real) Hilbert space with inner product

$$\langle f, g \rangle_{H^m(U)} = \sum_{|\alpha| \le m} \langle D^\alpha f, D^\alpha g \rangle_{L^2(U)},$$

and we denote it by $H^m(U)$.

The completion of $C_c^m(U)$ with respect to the norm $\| \cdot \|_{H^m}$ is the closed subspace, usually denoted $H_0^m(U)$, of $H^m(U)$. This latter space is very convenient for handling functions with support in U that vanish on the boundary, as in (2.3.5).

The rest of this section can be ignored by readers who just want to use these spaces to study the last part of Chapter 2.

Define the Fourier transform $\widehat{f}$ of $f \in C_c(\mathbb{R}^d)$ by

$$\widehat{f}(y) = (2\pi)^{-d/2} \int_{\mathbb{R}^d} f(x) e^{-ix \cdot y} dx,$$

for all $y \in \mathbb{R}^d$. We take for granted the following basic properties, as discussed in Appendix B.

190

- $\widehat{D^\alpha f}(y) = |y|^\alpha \widehat{f}(y)$.
- (Plancherel theorem) $||\widehat{f}||_2 = ||f||_2$.
- (Fourier inversion). For all $x \in \mathbb{R}^d$, $f(x) = (2\pi)^{-d/2} \int_{\mathbb{R}^d} \widehat{f}(y)e^{ix \cdot y}dy$.

Using these, we find that for $f \in C^m(\mathbb{R}^d)$,

$$||f||^2_{H^m(U)} = \int_{\mathbb{R}^d} \sum_{|\alpha| \le m} |y|^{2\alpha}|\widehat{f}(y)|^2 dy,$$

and it is then easy to see that an equivalent norm to $|| \cdot ||_{H^m(U)}$ is given by

$$||f||'_{H^m(U)} := \left(\int_{\mathbb{R}^d} (1 + |y|^2)^m |\widehat{f}(y)|^2 dy \right)^{\frac{1}{2}}.$$

Furthermore, the norm $|| \cdot ||'_{H^m(U)}$ has meaning for all non-negative real numbers m, and so we may generalise the definition given above to define $H^m(U)$ and $H^m_0(U)$ for all such m to be the completion of $C^m(U)$ (respectively $C^m_0(U)$) with respect to the norm $|| \cdot ||'_{H^m(U)}$. From now on we will simply identify our two equivalent norms and write $|| \cdot ||_{H^m(U)}$ instead of $|| \cdot ||'_{H^m(U)}$. It is easy to see that $H^m(U) \subseteq H^p(U)$ whenever $m > p$ and that the embedding is continuous.

We present (without full proof) two important theorems about Sobolev spaces. The second of these plays a role in Chapter 6.

Theorem C.0.1 (Sobolev Embedding Theorem) *If $m \in \mathbb{N}$, then $H^p_0(U) \subseteq C^m(\overline{U})$ for all positive real numbers $p > m + d/2$.*

Proof. We just deal with the case $m = 0$ here. By Fourier inversion and the Cauchy–Schwarz inequality, for all $f \in C_c(U), x \in U$,

$$|f(x)| \le (2\pi)^{-d/2} \left| \int_{\mathbb{R}^d} e^{ix \cdot y} \widehat{f}(y)(1 + |y|^2)^{p/2}(1 + |y|^2)^{-p/2} dy \right|$$

$$\le (2\pi)^{-d/2} \left(\int_{\mathbb{R}^d} (1 + |y|^2)^p |\widehat{f}(y)|^2 dy \right)^{1/2} \left(\int_{\mathbb{R}^d} (1 + |y|^2)^{-p} dy \right)^{1/2}$$

$$\le C||f||_{H^p_0(U)},$$

where $C > 0$. Hence we have

$$||f||_\infty \le C||f||_{H^p_0(U)}. \tag{C.0.1}$$

To see where the first inequality came from, observe that

$$\int_{\mathbb{R}^d} (1 + |y|^2)^{-p} dy = (2\pi)^d \int_0^\infty \frac{r^{d-1}}{(1+r^2)^p} dr,$$

and if we write $p = d/2 + \alpha$ where $\alpha > 0$, then

$$\int_1^\infty \frac{r^{d-1}}{(1+r^2)^p} dr \leq \int_1^\infty r^{-1-2\alpha} dr < \infty.$$

Now let $f \in H_0^p(U)$. Then it is the limit in the Sobolev norm of a sequence (f_n) in $C_0^p(U)$. But then by (C.0.1) and standard arguments, f is the uniform limit of the sequence (f_n), and so it is continuous. $\qquad\square$

Theorem C.0.2 (Rellich–Kondrachov Theorem) *If s, $p \geq 0$ with $s \leq p$, then the embedding from $H^p(U)$ into $H^s(U)$ is compact.*

Proof. There is much about compact operators in Chapter 5. For now it is enough to recall the definition, which in this context, requires us to establish that if (f_n) is a bounded sequence in $H_0^p(U)$, so that there exists $K > 0$ with $\|f_n\|_{H_0^p(U)} \leq K$ for all $n \in \mathbb{N}$, then it has a subsequence which converges in $H^s(U)$. We only sketch the proof. In fact we omit a key step which is to show, using the Arzela–Ascoli theorem, that the sequence $(\widehat{f_n})$ has a subsequence that is uniformly convergent on compact intervals on $\mathbb{R}^d$. We abuse notation (in a standard way) and just write this subsequence as $(\widehat{f_r})$. We now show that the corresponding subsequence (f_r) is Cauchy in $H^s(U)$, from which the result follows. Taking m and n sufficiently large we have for any $R > 0$

$$\|f_m - f_n\|_{H^s(U)} = \int_{\mathbb{R}^d} |\widehat{f_m}(y) - \widehat{f_n}(y)|^2 (1 + |y|^2)^s dy$$

$$= \int_{|y| \leq R} |\widehat{f_m}(y) - \widehat{f_n}(y)|^2 (1 + |y|^2)^s dy$$

$$+ \int_{|y| > R} |\widehat{f_m}(y) - \widehat{f_n}(y)|^2 (1 + |y|^2)^s dy.$$

The first integral is majorised by $C \sup_{|y| \leq R} |\widehat{f_m}(y) - \widehat{f_n}(y)|^2$ with $C > 0$, which can be made arbitrarily small, as we have already discussed. For the second integral, we note that if $|y| > R$, then

$$(1 + |y|^2)^s (1 + R^2)^{p-s} \leq (1 + |y|^2)^p,$$

and so

$$\int_{|y| > R} |\widehat{f_m}(y) - \widehat{f_n}(y)|^2 (1 + |y|^2)^s dy$$

$$\leq \frac{1}{(1 + R^2)^{p-s}} \int_{|y| > R} |\widehat{f_m}(y) - \widehat{f_n}(y)|^2 (1 + |y|^2)^p dy$$

$$= \frac{1}{(1 + R^2)^{p-s}} \|f_n - f_m\|^2_{H^p(U)}$$

$$\leq \frac{4K^2}{(1 + R^2)^{p-s}},$$

which can be made arbitrarily small by taking R to be sufficiently large, and the required result follows. $\qquad\square$

Appendix D

Probability Measures and Kolmogorov's Theorem on Construction of Stochastic Processes

Recall that a *probability space* $(\Omega, \mathcal{F}, P)$ is a measure space in which the finite measure P has total mass 1, i.e., $P(\Omega) = 1$. Let $(E, \mathcal{E})$ be a measurable space. A *random variable* X that takes values in E is defined to be a measurable function from $(\Omega, \mathcal{F})$ to $(E, \mathcal{E})$, i.e., X is a mapping from Ω to E so that $X^{-1}(A) \in \mathcal{F}$ for all $A \in \mathcal{E}$. The *law* or distribution of the random variable X is the probability measure p_X on $(E, \mathcal{E})$ which is the pushforward of P by X, so that for all $A \in \mathcal{E}$,

$$p_X(A) = P(X^{-1}(A)).$$

The standard examples that occur in many applications are real-valued random variables, where $(E, \mathcal{E}) = (\mathbb{R}, \mathcal{B}(\mathbb{R}))$, and real-valued random vectors, where $(E, \mathcal{E}) = (\mathbb{R}^d, \mathcal{B}(\mathbb{R}^d))$. A *stochastic process* is a family of random variables $X := (X_i, i \in \mathbb{I})$, indexed by some set $\mathbb{I}$ and taking values in $(E, \mathcal{E})$, which are all defined on the same probability space. In general, there is no reason why we should think of elements of $\mathbb{I}$ as representing "time", but in many important cases of interest, it does. For *discrete-time* stochastic processes, we typically take $\mathbb{I}$ to be $\mathbb{N}$, $\mathbb{Z}_+$ or $\mathbb{Z}$, while *continuous-time* stochastic processes may be indexed by $[0, T]$, $[0, \infty)$ or $\mathbb{R}$. The notion of law is then generalised to a collection of measures called the *finite-dimensional distributions* of the process X. These are defined as follows. For each m-tuple $(i_1, i_2, \ldots, i_m) \in \mathbb{I}^m$, where $m \in \mathbb{N}$ is arbitrary and $i_r \neq i_s$ for $r \neq s$, we obtain a measure $p_{i_1, i_2, \ldots, i_m}$ on $(E^m, \mathcal{E}^m)$ by

$$p_{i_1, i_2, \ldots, i_m}(A) = P(X^{-1}_{i_1, i_2, \ldots, i_m}(A)),$$

where $A \in \mathcal{E}^m$, and $X_{i_1, i_2, \ldots, i_m}$ is the random variable (random vector) defined by

$$X_{i_1, i_2, \ldots, i_m}(\omega) := (X_{i_1}(\omega), X_{i_2}(\omega), \ldots, X_{i_m}(\omega)),$$

for each $\omega \in \Omega$. Note that this family of measures satisfies two key conditions:

1. For all $\pi \in \Sigma(m)$, $A_1, A_2, \ldots, A_m \in \mathcal{E}$,

$$p_{i_{\pi(1)}, i_{\pi(2)}, \ldots, i_{\pi(m)}}(A_{\pi(1)} \times A_{\pi(2)} \cdots \times A_{\pi(m)}) = p_{i_1, i_2, \ldots, i_m}(A_1 \times A_2 \times \cdots \times A_m).$$

2. For all $B \in \mathcal{E}^m$,

$$p_{i_1, i_2, \ldots, i_m, i_{m+1}}(B \times E) = p_{i_1, i_2, \ldots, i_m}(B),$$

where $\Sigma(m)$ is the symmetric group on m letters. (1) and (2) are called the *Kolmogorov consistency conditions*. The key theorem due to Kolmogorov is that given any family of measures that satisfies (1) and (2), then there exists a stochastic process for which this family is the set of all the finite-dimensional distributions. For a full proof of this, see, e.g., Kallenberg [58], pp. 114–7 or Chapter 36 of Billingsley [12], pp. 430–8 for the case $E = \mathbb{R}$. We will not give the full proof here, but will try to say enough about the construction to give some insight to readers who have no prior knowledge of it.

So let us suppose that we are now given a family of probability measures $q_{i_1, i_2, \ldots, i_m}$ on $(E^m, \mathcal{E}^m)$ for $m \in \mathbb{N}$, $i_1, i_2, \ldots, i_m \in \mathbb{I}$, where $i_r \neq i_s$ for $r \neq s$, and assume that (1) and (2) are satisfied. We need to construct a probability space $(\Omega, \mathcal{F}, P)$ and a stochastic process $(X_i, i \in \mathbb{I})$ for which this given family of measures are the finite-dimensional distributions. We will take $\Omega = E^{\mathbb{I}}$, which is the set of all mappings from $\mathbb{I}$ to E, and our process X will be given by the "co-ordinate mappings", $X_i(\omega) = \omega(i)$, for $\omega \in \Omega, i \in \mathbb{I}$. Next we show how to construct $\mathcal{F}$. The building blocks for this σ-algebra are the *cylinder sets*, which are obtained as follows. Let $m \in \mathbb{N}, i_1, i_2, \ldots, i_n$ be distinct points in $\mathbb{I}$, and $B \in \mathcal{E}^m$. Define

$$A_B^{i_1, i_2, \ldots, i_m} := \{\omega \in \Omega : (\omega(i_1), \omega(i_2), \ldots \omega(i_m)) \in B\},$$

and let $\mathcal{C}(E, \mathbb{I})$ be the collection of all such cylinder sets. Then $\mathcal{C}(E, \mathbb{I})$ is an algebra of sets. Indeed we have $(A_B^{i_1, i_2, \ldots, i_m})^c = A_{B^c}^{i_1, i_2, \ldots, i_m}$, and at least when $\{i_1 i_2, \ldots i_m\} \cap \{j_1, j_2, \ldots, j_n\} = \emptyset$, we have

$$A_B^{i_1, i_2, \ldots, i_m} \cup A_C^{j_1, j_2, \ldots, j_m} = A_{B \cup C}^{i_1, i_2, \ldots, i_m, j_1, j_2, \ldots, j_n}.$$

The case where indices overlap is dealt with similarly but is a little more complicated to write down notationally. We then take $\mathcal{F}$ to be the σ-algebra generated by $\mathcal{C}(E, \mathbb{I})$.

Finally we must construct the measure P. We first define this on cylinder sets by, for $I = \{i_1, i_2, \ldots i_m\}$,

$$P(A_B^I) = q_{i_1, i_2, \ldots, i_m}(B).$$

Before we go any further, note that if P is indeed a measure, then from the construction of cylinder sets, and the prescription for the process X, we immediately have

$$q_{i_1,i_2,\ldots,i_m}(B) = P((X_{i_1}, X_{i_2}, \ldots, X_{i_m}) \in B),$$

as was required. Let us at least show that P is finitely additive. Using an obvious notation, we have for disjoint A_B^I and A_C^J,

$$
\begin{aligned}
P(A_B^I \cup A_C^J) &= P(A_{B\cup C}^{I\cup J}) \\
&= q_{I\cup J}(B \cup C) \\
&= q_{I\cup J}(B) + q_{I\cup J}(C) \\
&= q_{I\cup J}(B \times E^n) + q_{I\cup J}(E^m \times C) \\
&= q_I(B) + q_J(C) \\
&= P(A_B^I) + P(A_C^J).
\end{aligned}
$$

The reader will have noted the need to use both consistency conditions (1) and (2) to make this work. We must go a little further to prove that the measure P is countably additive on $\mathcal{C}(E, \mathbb{I})$, after which a well-known theorem can be used to show that it extends to a measure on $(\Omega, \mathcal{F})$.

Remark Although Kolomogorov's theorem is very powerful, it tells us nothing about sample path regularity. So, for example, it cannot be used on its own to construct Brownian motion from the heat kernel, as it will not deliver the required continuity of sample paths. That requires some further work (see, e.g., section 37 of Billingsley [12], pp. 444–7).

Appendix E

Absolute Continuity, Conditional Expectation and Martingales

E.1 The Radon–Nikodym Theorem

Let (S, Σ, μ) be a measure space. Recall that μ is *finite* if $\mu(S) < \infty$ and σ-*finite* if there exists a sequence $(A_n, n \in \mathbb{N})$ of mutually disjoint sets in Σ, such that $\bigcup_{n \in \mathbb{N}} A_n = S$ and $\mu(A_n) < \infty$ for all $n \in \mathbb{N}$.

Let μ and ν be two distinct measures defined on (S, Σ). We say that ν is *absolutely continuous* with respect to μ if, whenever $B \in \Sigma$ with $\mu(B) = 0$, then $\nu(B) = 0$. The measures μ and ν are said to be *equivalent* if they are mutually absolutely continuous. The key result on absolutely continuous measures is the Radon–Nikodym theorem, which we will prove below. Two proofs of this celebrated result may be found in Bogachev [15]. The first of these is in Chapter 3, section 3.2, pp. 178–9 and is essentially measure theoretic. The second is more functional analytic, and may be found in Chapter 4, section 4.3, pp. 256–7, and we present it below. The key tool we need from the theory of Hilbert spaces is

Theorem E.1.1 (The Riesz Representation Theorem) *If l is a bounded linear functional defined on a real Hilbert space H, then there exists a unique $y \in H$ such that*

$$l(x) = \langle x, y \rangle,$$

for all $x \in H$.

For a proof, see, e.g., Theorem II.4 on p. 43 of Reed and Simon [76].

Theorem E.1.2 (The Radon–Nikodym Theorem) *If μ and ν are measures defined on a measurable space (S, Σ), wherein ν is finite, μ is σ-finite and ν is* *absolutely continuous with respect to μ, then there exists a unique $f \in L^1(\mu)$ such that for all $A \in \Sigma$,*

$$v(A) = \int_A f(x)\mu(dx).$$

In particular, any version of f is non-negative (μ a.e.).

Proof. (Existence) First assume that μ and v are both finite measures. Define another finite measure, $\lambda := \mu + v$. It is easy to check that if a measurable function $h: S \to \mathbb{R}$ is λ-integrable, then it is μ-integrable. The mapping $L: L^2(\lambda) \to \mathbb{R}$, given for each $h \in L^2(\lambda)$, by

$$L(h) := \int_S h(x)\mu(dx),$$

is then easily seen to be well-defined and linear. By the Cauchy–Schwarz inequality, we have

$$|L(h)| \le \mu(S)^{\frac{1}{2}}||h||_\mu \le \mu(S)^{\frac{1}{2}}||h||_\lambda,$$

and so l is a bounded linear functional. Hence by the Riesz representation theorem (Theorem E.1.1), there exists a unique $g \in L^2(\lambda)$ such that

$$L(h) = \langle h, g \rangle_\lambda = \int_S h(x)g(x)\lambda(dx).$$

But then, by the definition of L, we see that we must have $d\mu = g d\lambda$, from which it follows that $dv = (1 - g)d\lambda$. Then our candidate to be f in the theorem is $(1 - g)/g$. To make this precise, first define $B := g^{-1}((-\infty, 0])$. Then $B \in \Sigma$ and

$$\mu(B) = \int_S \mathbf{1}_B(x)g(x)\lambda(dx) \le 0,$$

and so $\mu(B) = 0$. So we conclude that $g > 0$ (a.e. μ). Next define $C := g^{-1}((1, \infty))$. Again we have $C \in \Sigma$, and

$$\mu(C) = \int_S \mathbf{1}_C(x)g(x)\lambda(dx) > \lambda(C).$$

But we know that $\mu(C) \le \lambda(C)$. Hence $\mu(C) = 0$ and $0 < g < 1$ (a.e. μ). Now define

$$f(x) = \begin{cases} \frac{1-g(x)}{g(x)} & \text{if } x \notin B, \\ 0 & \text{if } x \in B. \end{cases}$$

Then f is well-defined, non-negative and measurable. It is also μ-integrable as

$$\int_S f(x)\mu(dx) = \int_{S\setminus B} f(x)\mu(dx)$$

$$= \int_{S\setminus B} (1 - g(x))\lambda(dx)$$

$$= v(S \setminus B) < \infty.$$

Now consider the sequence of measurable functions $(\mathbf{1}_{\{g\geq 1/n\}}, n \in \mathbb{N})$ which increases pointwise to 1 (a.e. μ). Hence, using absolute continuity of v with respect to μ, for each non-negative measurable function $H\colon S \to \mathbb{R}$, the sequence $(H\mathbf{1}_{\{g\geq 1/n\}}, n \in \mathbb{N})$ increases pointwise to H (a.e. v). So in particular, we have that for each $A \in \Sigma$, the sequence $(\mathbf{1}_A\mathbf{1}_{\{g\geq 1/n\}}, n \in \mathbb{N})$ increases pointwise to $\mathbf{1}_A$ (a.e. v), and $(f\mathbf{1}_{\{g\geq 1/n\}}, n \in \mathbb{N})$ which increases pointwise to f (a.e. v). Now using the monotone convergence theorem (twice), we obtain

$$v(A) = \lim_{n\to\infty} \int_S \mathbf{1}_A(x)\mathbf{1}_{\{g\geq 1/n\}}(x)v(dx)$$

$$= \lim_{n\to\infty} \int_S \mathbf{1}_A(x)\mathbf{1}_{\{g\geq 1/n\}}(x)f(x)\mu(dx)$$

$$= \int_A f(x)\mu(dx),$$

as required. Clearly we can modify f on any set of μ measure zero, and the result continues to hold.

The extension to σ-finite μ is carried out by writing $S = \bigcup_{n\in\mathbb{N}} A_n$, where $\mu(A_n) < \infty$ for all $n \in \mathbb{N}$. Then the result just proved shows that there exists f_n such that $v(C) = \int_C f_n(x)\mu(dx)$, for all $C \in \Sigma$ with $C \subseteq A_n$. Then the required function is obtained by defining

$$f = \sum_{n=1}^{\infty} f_n\mathbf{1}_{A_n}.$$

The details are left to the reader.

(Uniqueness) Assume there exist two distinct μ-integrable functions f_1 and f_2 such that for all $A \in \Sigma$,

$$v(A) = \int_A f_1(x)d\mu(x) = \int_A f_2(x)\mu(dx).$$

So $\int_A (f_1(x) - f_2(x))\mu(dx) = 0$. Now take $A = (f_1 - f_2)^{-1}([0, \infty)) \in \Sigma$, and observe that

$$\int_S |f_1(x) - f_2(x)|\mu(dx) = \int_A (f_1(x) - f_2(x))\mu(dx)$$

$$+ \int_{A^c} (f_2(x) - f_1(x))\mu(dx) = 0.$$

Hence $f_1 = f_2$ (a.e. μ) as required. $\qquad\square$

Note that the statement of the Radon–Nikodym theorem may be extended to the case where the measure v is also σ-finite. In that case the function f that appears in the statement of the theorem may no longer be integrable, but it must, of course, be measurable.

If the conditions of the Radon–Nikodym theorem are satisfied, then any version of the equivalence class of functions f that appears there is called a *Radon–Nikodym derivative* of v with respect to μ. All such versions agree, except on a set of μ measure zero. It is common to employ the notation $\frac{d\mu}{dv}$ for either f or one of its versions (the precise meaning is usually clear from the context).

We will look at two important applications of the Radon–Nikodym theorem to probability theory. The first of these concerns a random variable X, defined on a probability space $(\Omega, \mathcal{F}, P)$ and taking values in $\mathbb{R}^d$. Recall that the law p_X of X is the probability measure defined on $(\mathbb{R}^d, \mathcal{B}(\mathbb{R}^d))$ by the prescription $p_X(A) = P(X^{-1}(A))$ for all $A \in \mathcal{B}(\mathbb{R}^d)$. Assume that p_X is absolutely continuous with respect to Lebesgue measure on $(\mathbb{R}^d, \mathcal{B}(\mathbb{R}^d))$. Any version of the corresponding Radon–Nikodym derivative is called a *density* of X and is denoted f_X. So for all $A \in \mathcal{B}(\mathbb{R}^d)$, we have:

$$P(X \in A) = p_X(A) = \int_A f_X(x)dx.$$

f_X is a *probability density function* in that $f_X \geq 0$ (a.e.) and $\int_{\mathbb{R}^d} f_X(x)dx = 1$.

The other application of the Radon–Nikodym theorem to probability theory deserves a section to itself.

E.2 Conditional Expectation

Let $(\Omega, \mathcal{F}, P)$ be a probability space and X be a random variable defined on that space. We assume that X is *integrable* with respect to the measure P, i.e.,

$$\mathbb{E}(|X|) = \int_\Omega |X(\omega)|dP(\omega) < \infty.$$

Now let $\mathcal{G}$ be a sub-σ-algebra of $\mathcal{F}$. Everything that we need in this section follows from the following theorem:

Theorem E.2.3 *There exists a $\mathcal{G}$-measurable random variable Y such that for all $A \in \mathcal{G}$,*

$$\mathbb{E}(Y\mathbf{1}_A) = \mathbb{E}(X\mathbf{1}_A). \qquad (\text{E.2.1})$$

Moreover Y is the unique such $\mathcal{G}$-measurable random variable, up to modification on a set of measure zero.

Proof. (Existence) First assume that $X \geq 0$ (a.s.). Then it is not difficult to check that the prescription $Q_X(A) = \mathbb{E}(X\mathbf{1}_A)$, for all $A \in \mathcal{G}$, defines a measure Q_X on $(\Omega, \mathcal{G})$. Clearly $P(A) = 0 \Rightarrow Q_X(A) = 0$, i.e., Q_X is absolutely continuous with respect to P. So by the Radon–Nikodym theorem (Theorem E.1.2), there exists a $\mathcal{G}$-measurable random variable Y such that for all $A \in \mathcal{G}$,

$$\mathbb{E}(X\mathbf{1}_A) = Q_X(A) = \int_A Y(\omega)dP(\omega) = \mathbb{E}(Y\mathbf{1}_A),$$

i.e., Y is the Radon–Nikodym derivative of Q_X with respect to P, and so is uniquely defined up to a set of P measure zero. The general case follows easily by writing $X = X_+ - X_-$, where $X_+ := \max\{X, 0\}$ and $X_- := \max\{-X, 0\}$.

(Uniqueness) This is very similar to the corresponding part of the proof of the Radon–Nikodym theorem. Let Y_1 and Y_2 be distinct $\mathcal{G}$-random variables satisfying (E.2.1). Then the events $(Y_1 - Y_2)^{-1}([0, \infty))$ and $(Y_1 - Y_2)^{-1}((-\infty, 0])$ are both in $\mathcal{G}$. Hence

$$\mathbb{E}((Y_1 - Y_2)\mathbf{1}_{\{Y_1-Y_2\geq 0\}}) = E(X\mathbf{1}_{\{Y_1-Y_2\geq 0\}}) - E(X\mathbf{1}_{\{Y_1-Y_2\geq 0\}}) = 0.$$

Similarly $\mathbb{E}((Y_2 - Y_1)\mathbf{1}_{Y_1-Y_2\leq 0}) = 0$, and so

$$\mathbb{E}(|Y_1 - Y_2|) = \mathbb{E}((Y_1 - Y_2)\mathbf{1}_{\{Y_1-Y_2\geq 0\}}) + \mathbb{E}((Y_2 - Y_1)\mathbf{1}_{\{Y_1-Y_2\leq 0\}}) = 0.$$

Hence $Y_1 = Y_2$ (a.s.), and the result follows. $\qquad\square$

Any version of the $\mathcal{G}$-measurable random variable Y defined by (E.2.1) is called the *conditional expectation of Y given $\mathcal{G}$*, and is usually denoted by $\mathbb{E}(X|\mathcal{G})$.

Conditional expectation satisfies a number of useful properties, many of which are easily derived from (E.2.1). We will just list some of the most useful ones below and leave proofs to the reader to either work out for themselves or find in a suitable reference, such as the excellent Williams [99]. In all of the following, X is an integrable random variable defined on $(\Omega, \mathcal{F}, P)$.

(CE1) *Linearity.* If Y is integrable and $a, b \in \mathbb{R}$,

$$\mathbb{E}(aX + bY|\mathcal{G}) = a\mathbb{E}(X|\mathcal{G}) + b\mathbb{E}(Y|\mathcal{G}).$$

(CE2) *Positivity.* If $X \geq 0$, then $\mathbb{E}(X|\mathcal{G}) \geq 0$.

(CE3) *Taking out what is known.* If Y is $\mathcal{G}$-measurable and XY is integrable,

$$\mathbb{E}(XY|\mathcal{G}) = Y\mathbb{E}(X|\mathcal{G}).$$

(CE4) *The tower property.* If $\mathcal{H} \subseteq \mathcal{G} \subseteq \mathcal{F}$, then

$$\mathbb{E}(\mathbb{E}(X|\mathcal{G})|\mathcal{H}) = \mathbb{E}(X|\mathcal{H}).$$

(CE5) *Consistency.*

$$\mathbb{E}(\mathbb{E}(X|\mathcal{G})) = \mathbb{E}(X).$$

(CE6) *Independence.* If X and $\mathcal{G}$ are independent, then

$$\mathbb{E}(X|\mathcal{G}) = \mathbb{E}(X).$$

(CE7) *Simple estimation.*

$$|\mathbb{E}(X|\mathcal{G})| \leq \mathbb{E}(|X||\mathcal{G}).$$

(CE8) *Conditional Jensen's inequality.* If $f : \mathbb{R} \to \mathbb{R}$ is convex and $f(X)$ is integrable, then

$$f(\mathbb{E}(X|\mathcal{G})) \leq \mathbb{E}(f(X)|\mathcal{G}).$$

Note that when we restrict the action of the conditional expectation mapping $\mathbb{E}(\cdot|\mathcal{G})$ to square-integrable random variables, i.e. to $L^2(\Omega, \mathcal{F}, P) \subset L^1(\Omega, \mathcal{F}, P)$, then it coincides with the orthogonal projection $P_{\mathcal{G}}$ from $L^2(\Omega, \mathcal{F}, P)$ to the closed subspace $L^2(\Omega, \mathcal{G}, P)$. To see this, it is enough to observe that for all $X \in L^2(\Omega, \mathcal{F}, P)$, $A \in \mathcal{G}$,

$$
\begin{aligned}
\langle P_{\mathcal{G}}(X), \mathbf{1}_A \rangle_{L^2(\mathcal{G})} &= \langle X, P_{\mathcal{G}}^*(\mathbf{1}_A) \rangle_{L^2(\mathcal{F})} \\
&= \langle X, \mathbf{1}_A \rangle_{L^2(\mathcal{F})} \\
&= \mathbb{E}(X\mathbf{1}_A) \\
&= \mathbb{E}(\mathbb{E}(X|\mathcal{G})\mathbf{1}_A) \\
&= \langle \mathbb{E}(X|\mathcal{G}), \mathbf{1}_A \rangle_{L^2(\mathcal{G})},
\end{aligned}
$$

where we have used (E.2.1). Hence for all $A \in \mathcal{G}$,

$$\langle P_{\mathcal{G}}(X) - \mathbb{E}(X|\mathcal{G}), \mathbf{1}_A \rangle_{L^2(\mathcal{G})} = 0,$$

and so $P_{\mathcal{G}}(X) - \mathbb{E}(X|\mathcal{G}) = 0$ (a.s.) as the set $\{\mathbf{1}_A, A \in \mathcal{G}\}$ is total[1] in $L^2(\Omega, \mathcal{G}, P)$.

---

[1] Recall that a set of vectors in a Hilbert space is *total* if its orthogonal complement is $\{0\}$; equivalently, the linear span of the set is dense in the given space.

E.3 Martingales

Let $(\Omega, \mathcal{F}, P)$ be a probability space. A (discrete-time) *filtration* of $\mathcal{F}$ is a family $(\mathcal{F}_n, n \in \mathbb{N})$ of sub-σ-algebras of $\mathcal{F}$ such that $\mathcal{F}_n \subseteq \mathcal{F}_{n+1}$ for all $n \in \mathbb{N}$. A discrete-time stochastic process $(M_n, n \in \mathbb{N})$ is said to be *adapted* to the filtration if M_n is $\mathcal{F}_n$-measurable for each $n \in \mathbb{N}$. We say that $(M_n, n \in \mathbb{N})$ is a *discrete parameter martingale* if it is adapted, integrable (i.e., $E(|M_n|) < \infty$ for all $n \in \mathbb{N}$), and it satisfies the martingale property:

$$\mathbb{E}(M_{n+1}|\mathcal{F}_n) = M_n,$$

for all $n \in \mathbb{N}$. To construct simple examples, take $(X_n, n \in \mathbb{N})$ to be a sequence of independent random variables all defined on $(\Omega, \mathcal{F})$. For each $n \in \mathbb{N}$, let $\mathcal{F}_n$ be the smallest sub-σ-algebra of $\mathcal{F}$ with respect to which $X_1, X_2, \ldots, X_n$ are all measurable.

- If $\mathbb{E}(X_n) = 0$ for all $n \in \mathbb{N}$, then $(M_n, n \in \mathbb{N})$ is a martingale, where

$$M_n = X_1 + X_2 + \cdots + X_n.$$

- If X_n is non-negative and $\mathbb{E}(X_n) = 1$ for all $n \in \mathbb{N}$, then $(M_n, n \in \mathbb{N})$ is a martingale, where

$$M_n = X_1.X_2.\cdots X_n.$$

In the first case, the martingale property is established using (CE1) and (CE6), and in the second case, (CE3) and (CE6).

Discrete-time martingales are very interesting, but in the main part of the book we need the continuous version. A (continuous-time) *filtration* of $\mathcal{F}$ is a family $(\mathcal{F}_t, t \geq 0)$ of sub-σ-algebras of $\mathcal{F}$ such that $\mathcal{F}_s \subseteq \mathcal{F}_t$ whenever $0 \leq s \leq t < \infty$. A continuous-time stochastic process $(X(t), t \geq 0)$ is *adapted* to the filtration if $X(t)$ is $\mathcal{F}_t$-measurable for each $t \geq 0$. We say that $(X(t), t \geq 0)$ is a *continuous parameter martingale* if it is adapted, integrable (i.e., $E(|X(t)|) < \infty$ for all $t \geq 0$), and it satisfies the martingale property:

$$\mathbb{E}(X(t)|\mathcal{F}_s) = X(s),$$

for all $0 \leq s \leq t < \infty$.

Many interesting continuous-time martingales can be constructed from Brownian motion $(B(t), t \geq 0)$. We recall the definition of this, as given in section 3.1.2, and for simplicity we just take $d = 1$. We take our filtration to be the *natural* one given by the process, i.e., for all $t \geq 0$, $\mathcal{F}_t$ is the smallest sub-σ-algebra of $\mathcal{F}$ with respect to which $B(u)$ is measurable for all $0 \leq u \leq t$. First we note that Brownian motion is itself a martingale. To verify the martingale property, for each $0 \leq s \leq t < \infty$ write

$$B(t) = B(s) + (B(t) - B(s)),$$

and note that by (CE6), $\mathbb{E}(B(t) - B(s))|\mathcal{F}_s) = 0$.

For a less trivial example of a martingale based on Brownian motion, define $M(t) = B(t)^2 - 2t$ for all $t \geq 0$. Then for $0 \leq s \leq t < \infty$, we have

$$
\begin{aligned}
B(t)^2 &= [B(s) + (B(t) - B(s))]^2 \\
&= B(s)^2 + 2B(s)(B(t) - B(s)) + (B(t) - B(s))^2.
\end{aligned}
$$

Using (CE3) and (CE6), we find that $E(B(s)(B(t) - B(s)))|\mathcal{F}_s) = 0$, and by (CE6) again, together with (B3), $\mathbb{E}(B(t) - B(s))^2|\mathcal{F}_s) = 2(t - s)$. The martingale property is easily deduced from here.

A large class of martingales may be constructed by using stochastic integration. This is described in the next appendix.

Appendix F

Stochastic Integration and Itô's Formula

F.1 Stochastic Integrals

The material in this appendix will not seek to present ideas in full generality, and will aim to display the guts of constructions and proofs without giving all the details. For further insight, see, e.g., Baudoin [10], Oksendal [71], Steele [92], Rogers and Williams [80], and references therein.

As in the previous appendix, we will restrict ourselves to consideration of a one-dimensional Brownian motion $(B(t), t \geq 0)$ defined on a probability space $(\Omega, \mathcal{F}, P)$. We equip this space with a filtration $(\mathcal{F}_t, t \geq 0)$, and we will strengthen the independent increments requirement (B1) of Chapter 3, and require that for all $0 \leq s \leq t < \infty$ the random variable $B(t) - B(s)$ is independent of the σ-algebra $\mathcal{F}_s$, i.e., for all $A \in \mathcal{B}(\mathbb{R})$ and $C \in \mathcal{F}_s$,

$$P((B(t) - B(s))^{-1}(A) \cap C) = P((B(t) - B(s))^{-1}(A))P(C).$$

This condition is equivalent to (B1) when we use the natural filtration of Brownian motion.

We aim to make sense of the *stochastic integral*,

$$I_T(F) = \int_0^T F(s)\, dB(s), \tag{F.1.1}$$

for $T > 0$, and so we will from now just work on the interval $[0, T]$. Since the paths of Brownian motion are almost surely non-differentiable and not even of finite variation on compact sets, there is no hope of making sense of (F.1.1) within a pathwise Riemann–Stieltjes or Lebesgue–Stieltjes framework. We require a new notion of integral, as introduced by Itô and Doeblin. For this to work, we introduce the real Hilbert space $\mathcal{H}_T^2$ of all adapted real-valued processes $(F(t), 0 \leq t \leq T)$ such that $\int_0^T \mathbb{E}(F(s)^2) ds < \infty$. The inner product in $\mathcal{H}_T^2$ is given by

$$\langle F, G \rangle_T = 2 \int_0^T \mathbb{E}(F(s)G(s)) ds.$$

It can be shown that Σ_T is dense in $\mathcal{H}_T^2$, where $F \in \Sigma_T$ if there exists a partition $0 = t_0 < t_1 < \cdots < t_{N+1} = T$ such that[1] for all $0 < t \leq T$,

$$F(t) = \sum_{j=0}^{N} F(t_j) \mathbf{1}_{(t_j, t_{j+1}]}(t).$$

For $F \in \Sigma_T$, we define its stochastic integral by

$$I_T(F) = \sum_{j=0}^{N} F(t_j)(B(t_{j+1}) - B(t_j)). \tag{F.1.2}$$

In order to extend the definition further, we must compute

$$\mathbb{E}(I_T(F)^2) = \sum_{j=0}^{N} \sum_{k=0}^{N} \mathbb{E}[F(t_j)F(t_k)(B(t_{j+1}) - B(t_j))(B(t_{k+1}) - B(t_k))].$$

We next split this double sum into three pieces, these being the sums over all $j < k$, $j > k$ and $j = k$, respectively. For the first of these, we have by (CE3), the stronger version of (B1) imposed above, (C6) and (B3) that

$$\sum_{j<k} \mathbb{E}[F(t_j)F(t_k)(B(t_{j+1}) - B(t_j))(B(t_{k+1}) - B(t_k))]$$

$$= \sum_{j<k} \mathbb{E}[\mathbb{E}(F(t_j)(B(t_{j+1}) - B(t_j))F(t_k)|\mathcal{F}_{t_k})\mathbb{E}(B(t_{k+1}) - B(t_k))]$$

$$= 0.$$

By a similar argument, the sum over $j > k$ also vanishes. We are then left with

$$\mathbb{E}(I_T(F)^2) = \sum_{j=0}^{N} \mathbb{E}(F(t_j)^2(B(t_{j+1}) - B(t_j))^2)$$

$$= \sum_{j=0}^{N} \mathbb{E}[\mathbb{E}(F(t_j)^2|\mathcal{F}_{t_j})\mathbb{E}((B(t_{j+1}) - B(t_j))^2)]$$

---

[1] There is a good reason why we set this up so that F is pathwise left continuous, but we will not go into that here.

$$= 2 \sum_{j=0}^{N} \mathbb{E}(F(t_j)^2)(t_{j+1} - t_j)$$

$$= 2 \int_0^T \mathbb{E}(F(t)^2)dt = ||F||_T^2.$$

So I_T is an isometry from Σ_T to $L^2(\Omega, \mathcal{F}, P)$, and hence it extends uniquely to an isometry from $\mathcal{H}_T^2$ to $L^2(\Omega, \mathcal{F}, P)$, which we continue to denote as I_T. This mapping I_T is called *Itô's isometry*, and $I_T(F)$, for $F \in \mathcal{H}_T^2$, is the stochastic integral of the process F. Note that the stochastic integral, so defined, is a random variable, but by varying $T > 0$ and imposing suitable conditions on T (which the reader may easily figure out), we may also consider the stochastic process $(I_t(F), t \geq 0)$ wherein $I_0(F) := 0$. We will also use the more suggestive integral notation for $I_T(F)$, as in (F.1.1), where appropriate.

The following results are all fairly straightforward to establish, or can be looked up in standard texts. In all of the following, $F \in \mathcal{H}_T^2$,

- For all $G \in \mathcal{H}_T^2$ and $\alpha, \beta \in \mathbb{R}$,

$$\int_0^T (\alpha F(t) + \beta G(t))dB(t) = \alpha \int_0^T F(t)dB(t) + \beta \int_0^T G(t)dB(t).$$

- $\mathbb{E}(I_T(F)) = 0, \mathbb{E}(I_T(F)^2) = ||F||_T^2$.
- The process $(I_t(F), t \geq 0)$ is adapted to the given filtration.
- The process $(I_t(F), t \geq 0)$ is a martingale.

To establish the last of these, first consider the case where $F \in \Sigma_T$, and compute $\mathbb{E}(I_t(F)|\mathcal{F}_s)$, taking s (without loss of generality) to be one of the points of the partition.

To make the stochastic integral into a more powerful tool, we may extend it to the case where $F \in \mathcal{P}_T^2$, which is the space of all adapted processes $F = (F(t), 0 \leq t \leq T)$ for which $P\left(\int_0^T F(s)^2 ds < \infty\right) = 1$. This is not a Hilbert space, and the extended stochastic integral is no longer a martingale – it is in fact a more general object called a *local martingale*. We will not pursue this theme further here.

Once we have a theory of stochastic integration, we can consider *Itô processes* $(M(t), t \geq 0)$ which take the form

$$M(t) = M(0) + \int_0^t F(s)dB(s) + \int_0^t G(s)ds, \qquad \text{(F.1.3)}$$

where $(G(t), t \geq 0)$ is an adapted process having (almost surely) finite variation on bounded intervals. Then the last integral may be defined as a pathwise Lebesgue integral. Such processes are very useful in mathematical modelling.

F.2 Itô's Formula

In this section we sketch the proof of the celebrated Itô formula for processes of the form $(f(B(t)), t \geq 0)$, where $f \in C^2(\mathbb{R})$. A key step along the way is the computation of a key quantity that is called the *quadratic variation*. We will not spend time here exploring that concept, but will simply present the required calculation within a lemma. First fix $T > 0$ and let $(\mathcal{P}_n, n \in \mathbb{N})$ be a sequence of partitions of $[0, T]$. We take $\mathcal{P}_n$ to be $0 = t_0^{(n)} < t_1^{(n)} < \cdots < t_{N_n+1}^{(n)} = T$ and write $(\delta_n, n \in \mathbb{N})$ for the associated sequence of meshes, so that for each $n \in \mathbb{N}$, $\delta_n := \max_{0 \leq j \leq N_n}(t_{j+1}^{(n)} - t_j^{(n)})$. Finally we assume that $\lim_{n \to \infty} \delta_n = 0$.

Lemma F.2.1 *If $G \in \Sigma_T$, then*

$$\lim_{n \to \infty} \mathbb{E}\left[\left(\sum_{j=0}^{N_n} G(t_j^{(n)})(B(t_{j+1}^{(n)}) - B(t_j^{(n)}))^2 - 2\int_0^T G(s)ds\right)^2\right] = 0.$$

Proof. To simplify the presentation of the following, we will drop all n subscripts and superscripts, and introduce the compact notation $G_j = G(t_j)$, $\delta t_j = t_{j+1} - t_j$ and $\Delta B_j = B(t_{j+1}) - B(t_j)$ for $j = 0, \ldots, N$. We are interested in the quantity $V = V_N$, where

$$V := \mathbb{E}\left[\left(\sum_{j=0}^{N} G_j \Delta B_j^2 - 2\sum_{j=0}^{N} G_j \Delta t_j\right)^2\right]$$

$$= \mathbb{E}\left[\sum_{j=0}^{N}\sum_{k=0}^{N} G_j(\Delta B_j^2 - 2\Delta t_j)G_k(\Delta B_k^2 - 2\Delta t_k)\right].$$

We again split the double sum into the sums over all $j < k$, $j > k$ and $j = k$, respectively. The reader is invited to check that the first two of these vanish. Then we have

$$V = \mathbb{E}\left(\sum_{j=0}^{N} G_j^2(\Delta B_j^2 - 2\Delta t_j)^2\right)$$

$$= \sum_{j=0}^{N} \mathbb{E}(G_j^2[\Delta B_j^4 - 4\Delta B_j^2 \Delta t_j + 4\Delta t_j^2]).$$

By well-known properties of moments of Gaussian random variables, we have $\mathbb{E}(\Delta B_j^4) = 6\Delta t_j^2$. Then

$$V = \sum_{j=0}^{N} \mathbb{E}(G_j^2)(6\Delta t_j^2 - 8\Delta t_j^2 + 4\Delta t_j^2)$$

$$= 2 \sum_{j=0}^{N} \mathbb{E}(G_j^2)\Delta t_j^2,$$

and so

$$\lim_{n\to\infty} V_n \le 2 \lim_{n\to\infty} \delta_n \sum_{j=0}^{N} \mathbb{E}(G_j^2)\Delta t_j = 0,$$

since $\lim_{n\to\infty} \sum_{j=0}^{N} \mathbb{E}(G_j^2)\Delta t_j = \int_0^T \mathbb{E}(G(s)^2)ds$, and the required result follows. $\qquad\qquad\square$

Now we are ready for the main event.

Theorem F.2.2 (Itô's Formula) *If $(B(t), t \ge 0)$ is a Brownian motion defined on some probability space, and $f \in C^2(\mathbb{R})$, then for all $t \ge 0$ (with probability one),*

$$f(B(t)) - f(0) = \int_0^t f'(B(s))dB(s) + \int_0^t f''(B(s))ds. \qquad (\text{F.2.4})$$

Proof. We just sketch the main points of this. As in the proof of Lemma F.2.1, we drop all superscripts from partitions. Fix $t > 0$. Using Taylor's theorem (pathwise), we have that for each $0 \le j \le N$, there exists a $\mathcal{F}_{t_{j+1}}$-measurable random variable X_j so that $|X_j - B_{t_j}| \le |B_{t_{j+1}} - B(t_j)|$, and

$$f(B(t)) - f(0) = \sum_{j=0}^{N} f(B(t_{j+1})) - f(B(t_j))$$

$$= \sum_{j=0}^{N} f'(B(t_j))(B(t_{j+1}) - B(t_j))$$

$$+ \frac{1}{2} \sum_{j=0}^{N} f''(X_j)(B(t_{j+1}) - B(t_j))^2$$

$$= \sum_{j=0}^{N} f'(B(t_j))(B(t_{j+1}) - B(t_j))$$

$$+\frac{1}{2}\sum_{j=0}^{N} f''(B(t_j))(B(t_{j+1}) - B(t_j))^2$$

$$+\frac{1}{2}\sum_{j=0}^{N}(f''(X_j) - f''(B(t_j)))(B(t_{j+1}) - B(t_j))^2.$$

Using the definition of stochastic integral in the first term, and Lemma F.2.1 in the second, we obtain convergence as required to the right-hand side of (F.2.4). For the final term, we have that

$$\max_{i \le j \le N} |f''(X_j) - f''(B(t_j))| \to 0,$$

by continuity, while $\sum_{j=0}^{N}(B(t_j))(B(t_{j+1}) - B(t_j))^2$ converges in L^2 to $2t$, by Lemma F.2.1. Then (taking the limit along a subsequence, if necessary) it follows that this final term converges almost surely to zero, and we are done. $\square$

We remind the reader that most books on the subject will include a factor of $1/2$ in front of the term involving the second derivative in (F.2.4). That is because they are using "probabilist's Brownian motion", where $B(t)$ has variance t, rather than the "analyst's Brownian motion", where that variance is doubled. As a swift application of Itô's lemma, we derive a result we used frequently in Chapter 4, namely the characteristic function of Brownian motion (the Fourier transform of the Gauss–Weierstrass kernel). To that end, take f in Theorem F.2.2 to be given by $f(x) = e^{iux}$ where $u \in \mathbb{R}$. Then you can check that by treating real and imaginary parts separately in Theorem F.2.2, you obtain, for each $t > 0$,

$$e^{iuB(t)} = iu \int_0^t e^{iuB(s)} dB(s) - u^2 \int_0^t e^{iuB(s)} ds.$$

Taking expectations of both sides, and using the fact that stochastic integrals have zero mean, we obtain by Fubini's theorem that

$$\mathbb{E}(e^{iuB(t)}) = 1 - u^2 \int_0^t \mathbb{E}(e^{iuB(s)}) ds,$$

from which we easily deduce (e.g., by differentiating both sides of the last identity and solving the resulting differential equation) that

$$\mathbb{E}(e^{iuB(t)}) = \int_{\mathbb{R}} e^{iuy} \gamma_t(y) dy = e^{-tu^2}.$$

There are many generalisations of Itô's formula, for example, if we take $(B(t), t \ge 0)$ to be d-dimensional standard Brownian motion, as was done in

Chapter 3, then for all $t > 0$,

$$f(B(t)) - f(0) = \int_0^t \nabla f(B(s)) \cdot dB(s) + \int_0^t \Delta f(B(s)) ds,$$

or if we replace Brownian motion by (for simplicity) a one-dimensional Itô process as in (F.1.3), we have

$$f(M(t)) - f(M(0)) = \int_0^t f'(M(s)) F(s) dB(s) + \int_0^t f'(M(s)) G(s) ds$$

$$+ \int_0^t f''(M(s)) F(s)^2 ds.$$

Further generalisations may be found in the literature.

Appendix G

Measures on Locally Compact Spaces – Some Brief Remarks

When we go beyond $\mathbb{R}^d$ to general topological spaces, a very desirable property is that the space be *locally compact*, i.e., every point has an open neighbourhood that has compact closure. So let X be such a locally compact space, and $\mathcal{B}(X)$ be the usual Borel σ-algebra, i.e., the smallest σ-algebra of subsets of X that contains all the open sets. Then $(X, \mathcal{B}(X))$ is a measurable space, and we may consider Borel measures μ that are defined on it in the usual way. In this case, it turns out to be very profitable to consider a subclass of measures that satisfy some nice properties that are automatically satisfied when X is $\mathbb{R}^d$, and μ is Lebesgue measure. In the general case, we say that μ is *outer regular* if, for all $A \in \mathcal{B}(X)$,

$$\mu(A) = \inf\{\mu(O); A \subset O, O \text{ is open in } X\}.$$

The measure μ is said to be *inner regular* if, for all $A \in \mathcal{B}(X)$,

$$\mu(A) = \sup\{\mu(C); C \subset A, C \text{ is compact in } X\}.$$

We say that μ is *regular* if it is both inner and outer regular, and also satisfies $\mu(C) < \infty$ for all compact sets C in X.

Now let μ be a regular Borel measure defined on $(X, \mathcal{B}(X))$ and define a linear functional l_μ on the space $C_c(X)$ by the prescription

$$l_\mu(f) = \int_X f(x)\mu(dx),$$

for all $f \in C_c(X)$. Note that the values of $l_\mu(f)$ are indeed finite, since by regularity,

$$|l_\mu(f)| \leq \int_X |f(x)|\mu(dx) \leq ||f||_\infty \mu(K) < \infty,$$

where $K := \text{supp}(\mu)$. The linear function is also *positive*, in the sense that if $f \geq 0$, then $l_\mu(f) \geq 0$. This result has a powerful converse which we will state below. It is the celebrated *Riesz lemma*.[1]

Theorem G.0.1 (Riesz Lemma) *Let X be locally compact and Hausdorff. If I is a positive linear functional on $C_c(X)$, then there is a unique regular Borel measure μ on $(X, \mathcal{B}(X))$ such that for all $f \in C_c(X)$,*

$$I(f) = \int_X f(x)\mu(dx).$$

The proof is too technical to go into here. We refer readers to standard texts, such as section 113 of Bogachev [16], pp. 113–7 or section 7.2 of Cohn [22], pp. 205–17. It requires a great deal of detailed measure theory, interplayed with topology. We just indicate where the measure μ comes from. Let U be an open set in X and let

$$\mathcal{C}(O) = \{f \in C_c(X); 0 \leq f \leq \mathbf{1}_U\}.$$

We first define a set function μ^* on open sets by the prescription

$$\mu^*(O) = \sup\{I(f), f \in \mathcal{C}(O)\},$$

and then extend to all subsets of X by the prescription

$$\mu^*(A) = \inf\{\mu^*(O), O \text{ is open and } A \subseteq O\}.$$

We then show that μ^* is an outer measure, and apply Carathéodory's theorem to show that it restricts to a Borel measure μ on $(X, \mathcal{B}(X))$ as required. Having so constructed μ, it is still necessary to prove that it is regular, and that it uniquely represents l as in Theorem G.0.1.

---

[1] This is sometimes called the "Riesz representation theorem", but in this book we have reserved that name for the result that characterises the dual of a Hilbert space, as described in Theorem E.1.1 in Appendix E.

References

[1] R. Abraham, J. E. Marsden, T. Ratiu, *Manifolds, Tensor Analysis and Applications*, Springer-Verlag: New York, Heidelberg (1988).

[2] R. Adams, J. Fournier, *Sobolev Spaces* (second edition), Academic Press (2003).

[3] N. I. Akhiezer, I. M. Glazman, *Theory of Linear Operators in Hilbert Space, Volume 1*, Pitman (1981).

[4] S. Albeverio, R. Høegh–Krohn, *Mathematical Theory of Feynman Path Integrals*, Lecture Notes in Mathematics Vol. **523**, Springer (1976).

[5] R. Alicki, M. Fannes, Dilations of quantum dynamical semigroups with classical Brownian motion, *Commun. Math. Phys.* **108**, 353–61 (1987).

[6] D. Applebaum, *Lévy Processes and Stochastic Calculus* (second edition), Cambridge University Press (2009).

[7] D. Applebaum, *Probability on Compact Lie Groups*, Springer (2014).

[8] D. Applebaum, Probabilistic trace and Poisson summation formulae on locally compact abelian groups, *Forum Math.* **29**, 501–17 (2017); Corrigendum, *Forum Math.* **29**, 1499–1500 (2017).

[9] D. Applebaum, R. Bañuelos, Probabilistic approach to fractional integrals and the Hardy–Littlewood–Sobolev inequality, in "Analytic Methods in Interdisciplinary Applications", Springer Proceedings in Mathematics and Statistics Vol. **116**, 17–40 (2014).

[10] F. Baudoin, *Diffusion Processes and Stochastic Calculus*, European Mathematical Society (2014).

[11] C. Berg, G. Forst, *Potential Theory on Locally Compact Abelian Groups*, Springer-Verlag (1975).

[12] P. Billingsley, *Probability and Measure* (second edition), Wiley (1986).

[13] P. Billingsley, *Convergence of Probability Measures* (second edition), Wiley (1999).

[14] N. H. Bingham, C. M. Goldie, J. L. Teugels, *Regular Variation*, Cambridge University Press (1987).

[15] V. I. Bogachev, *Measure Theory, Volume 1*, Springer-Verlag (2007).

[16] V. I. Bogachev, *Measure Theory, Volume 2*, Springer-Verlag (2007).

[17] B. Böttcher, R. Schilling, J. Wang, *Lévy Matters III, Lévy Type Processes, Construction, Approximation and Sample Path Properties*, Lecture Notes in Mathematics Vol. **2099**, Springer International Publishing (2013).

[18] O. Bratteli, *Derivations, Dissipations and Group Actions on C*-Algebras*, Lecture Notes in Mathematics Vol. **1229**, Springer-Verlag (1986).

[19] O. Bratteli, D. W. Robinson, *Operator Algebras and Quantum Statistical Mechanics I*, Springer-Verlag (1979).

[20] O. Bratteli, D. W. Robinson, *Operator Algebras and Quantum Statistical Mechanics II*, Springer-Verlag (1981).

[21] E. A. Carlen, S. Kusuoka, D. W. Stroock, Upper bounds for symmetric Markov transition functions, *Ann. Inst. Henri Poincaré (Prob. Stat.)* **23**, 245–87 (1987).

[22] D. L. Cohn, *Measure Theory*, Birkhaüser (1980).

[23] P. Courrège, Sur la forme intégro-différentielle des opérateurs de C_k^∞ dans C satifaisant au principe du maximum, *Sém. Théorie du Potential* exposé **2**, (1965/66) 38 pp.

[24] G. Da Prato, J. Zabczyk, *Ergodicity for Infinite Dimensional Systems*, Cambridge University Press (1996).

[25] E. B. Davies, *One-Parameter Semigroups*, Academic Press (1980).

[26] E. B. Davies, *Heat Kernels and Spectral Theory*, Cambridge University Press (1989).

[27] E. B. Davies, *Linear Operators and their Spectra*, Cambridge University Press (2007).

[28] K. Deimling, *Ordinary Differential Equations in Banach Spaces*, Lecture Notes in Mathematics Vol. **596**, Springer-Verlag (1977).

[29] P. A. M.Dirac, *The Principles of Quantum Mechanics* (fourth edition), Oxford University Press (1958).

[30] R. M. Dudley, *Real Analysis and Probability*, Wadsworth and Brooks/Cole Advanced Books and Software (1989); republished, Cambridge University Press (2002).

[31] K-J. Engel, R. Nagel, *One–Parameter Semigroups for Linear Evolution Equations*, Springer-Verlag (2000).

[32] K-J. Engel, R. Nagel, *A Short Course on Operator Semigroups*, Springer (2006).

[33] S. N. Ethier, T. G. Kurtz, *Markov Processes, Characterisation and Convergence*, Wiley (1986).

[34] L. C. Evans, *Partial Differential Equations* (second edition), Graduate Studies in Mathematics Vol. 19, American Mathematical Society (2010).

[35] W. Feller, *An Introduction to Probability Theory and its Applications*, vol. 2 (second edition), Wiley (1971).

[36] P. Friz, Heat kernels, parabolic PDEs and diffusion processes, www.math .nyu.edu/ frizpete/publications/heatkernel/hk.pdf

[37] C. Foias, B. Sz-Nagy, *Harmonic Analysis of Operators on Hilbert Space*, North-Holland Amsterdam (1970).

[38] M. Fukushima, Y. Oshima, M. Takeda, *Dirichlet Forms and Symmetric Markov Processes*, de Gruyter (1994).

[39] H. Goldstein, *Classical Mechanics*, Addison-Wesley (1950).

[40] J. A. Goldstein, *Semigroups of Linear Operators and Applications*, Oxford University Press (1985), second edition Dover (2017).

[41]   A. Grigor'yan, *Heat Kernel and Analysis on Manifolds*, AMS/IP Studies in Advanced Mathematics Vol. 47, American Mathematical Society (2009).

[42]   A. Grigor'yan, A. Telcs, Two-sided estimates of heat kernels on metric measure spaces, *Annals of Probability* **40**, 1212–1284 (2012).

[43]   K. Hannabuss, *An Introduction to Quantum Theory*, Clarendon Press, Oxford (1997).

[44]   H. Heyer, *Probability Measures on Locally Compact Groups*, Springer-Verlag (1977).

[45]   E. Hille, *Functional Analysis and Semigroups*, American Math. Soc. (1948).

[46]   E. Hille, R. Phillips, *Functional Analysis and Semigroups*, American Math. Soc. (1957).

[47]   W. Hoh, *Pseudo Differential Operators Generating Markov Processes*, Habilitationsschrift Universität Bielefeld (1998), available from http://citeseerx.ist .psu.edu/viewdoc/download?doi=10.1.1.465.4876&rep=rep1&type=pdf

[48]   W. Hoh, The martingale problem for a class of pseudo-differential operators, *Math. Ann.* **300**, 121–47 (1994).

[49]   L. Hörmander, *The Analysis of Linear Partial Differential Operators I: Distribution Theory and Fourier Analysis* (second edition), Springer-Verlag (2003).

[50]   G. A. Hunt, Semigroups of measures on Lie groups, *Trans. Amer. Math. Soc.* **81**, 264–93 (1956).

[51]   K. Itô, *An Introduction to Probability Theory*, Cambridge University Press (1984).

[52]   N. Jacob, A class of Feller semigroups generated by pseudo differential operators, *Math Z.* **215**, 151–66 (1994).

[53]   N. Jacob, *Pseudo-Differential Operators and Markov Processes*, Akademie-Verlag, Mathematical Research 94 (1996).

[54]   N. Jacob, *Pseudo-Differential Operators and Markov Processes: 1, Fourier Analysis and Semigroups*, World Scientific (2001).

[55]   N. Jacob, *Pseudo-Differential Operators and Markov Processes: 2, Generators and Their Potential Theory*, World Scientific (2002).

[56]   N. Jacob, *Pseudo-Differential Operators and Markov Processes: 3, Markov Processes and Applications*, World Scientific (2005).

[57]   M. Kac, On the average of a certain Wiener functional and a related limit theorem in calculus of probability, *Trans. Amer. Math. Soc.* **59** 401–414 (1946).

[58]   O. Kallenberg, *Foundations of Modern Probability*, Springer-Verlag, (1997) (second edition) (2002).

[59]   T. Kato, *Perturbation Theory for Linear Operators* (second edition), Springer-Verlag (1995).

[60]   V. N. Kolokoltsov, *Markov Processes, Semigroups and Generators*, De Gruyter Studies in Mathematics **38**, Walter de Gruyter (2011).

[61]   A. Kyprianou, *Introductory Lectures on Fluctuations of Lévy Processes with Applications*, Springer-Verlag (2006).

[62]   A. Lasota, M. C. Mackey, *Chaos, Fractal and Noise, Stochastic Aspects of Dynamics*, (second edition) Springer-Verlag (1994).

[63]   J. M. Lindsay, *Quantum Stochastic Analysis – An Introduction*, in *Quantum Independent Increment Processes I* ed. M. Schürmann, U. Franz, pp. 181–273, Lecture Notes in Mathematics Vol. **1865** Sprnger-Verlag (2005).

[64] G. W. Mackey, *Mathematical Foundations of Quantum Mechanics*, W. A. Benjamin Inc. (1963), reprinted by Dover (2004).

[65] P. Malliavin (with H. Airault, L. Kay, G. Letac), *Integration and Probability*, Springer-Verlag (1995).

[66] M. M. Meerchaert, A. Sikorski, *Stochastic Models for Fractional Calculus*, De Gruyter Studies in Mathematics **43**, Walter de Gruyter (2012).

[67] A. J. Milani, N. J. Koksch, *An Introduction to Semiflows*, Chapman and Hall/CRC Press (2004).

[68] B. Misra, I. Prigogine, M. Courbage, From deterministic dynamics to probabilistic descriptions, *Physica* **98A**, 1–26 (1979).

[69] P. Mörters, Y. Peres, *Brownian Motion*, Cambridge University Press (2010).

[70] J. Nash, Continuity of solutions of parabolic and elliptic equations, *American J. Math.* **80**, 931–54 (1958).

[71] B. Øksendal, *Stochastic Differential Equations* (sixth edition), Springer-Verlag (2003).

[72] K. B. Oldham, J. Spanier, *The Fractional Calculus: Theory and Applications*, Academic Press (1974), Dover (2006).

[73] K. R. Parthasarathy, *An Introduction to Quantum Stochastic Calculus*, Birkhaüser Verlag (1992).

[74] A. Pazy, *Semigroups of Linear Operators and Applications to Partial Differential Equations*, Springer-Verlag (1983).

[75] I. Prigogine, I. Stengers, *Order Out of Chaos, Man's New Dialogue with Nature*, Heinemann Ltd. (1984).

[76] M. Reed, B. Simon, *Methods of Modern Mathematical Physics*, vol. 1, *Functional Analysis* (revised and enlarged edition), Academic Press (1980).

[77] M. Reed, B. Simon, *Methods of Modern Mathematical Physics*, vol. 2, *Fourier Analysis, Self-Adjointness*, Academic Press (1975).

[78] M. Renardy, R. C. Rogers, *An Introduction to Partial Differential Equations*, Springer-Verlag (1992).

[79] L. C. G. Rogers, D. Williams, *Diffusions, Markov Processes and Martingales*, vol. 1, *Foundations*, Wiley (1979, 1994); Cambridge University Press (2000).

[80] L. C. G. Rogers, D. Williams, *Diffusions, Markov Processes and Martingales*, vol. 2, *Itô Calculus*, Wiley (1994); Cambridge University Press (2000).

[81] S. Rosenberg, *The Laplacian on a Riemannian Manifold*, Cambridge University Press (1997).

[82] W. Rudin *Functional Analysis* (second edition), McGraw-Hill (1991).

[83] M. Ruzhansky, V. Turunen, *Pseudo-differential Operators and Symmetries: Background Analysis and Advanced Topics*, Birkhäuser (2010)

[84] K.-I. Sato, *Lévy Processes and Infinite Divisibility*, Cambridge University Press (1999).

[85] L. Saloff-Coste, *Aspects of Sobolev-Type Inequalities*, Cambridge University Press (2002).

[86] R. L. Schilling, An introduction to Lévy and Feller processes. In *From Lévy-type Processes to Parabolic SPDEs*, 1–126, Adv. Courses Math. CRM Barcelona, ed. L. Quer-Sardanyons, F. Utzet, Birkhauser/Springer (2016).

[87] R. L. Schilling, L. Partzsch, *Brownian Motion*, de Gruyter (2012).

[88]   R. L. Schilling, R. Song, Z. Vondraček, *Bernstein Functions, Theory and Applications*, Studies in Mathematics **37**, De Gruyter (2010).

[89]   G. F. Simmons, *Introduction to Topology and Modern Analysis*, McGraw Hill Book Company Inc (1963).

[90]   B. Simon, *Real Analysis, a Comprehensive Course in Analysis, Part 1*, American Mathematical Society (2015).

[91]   K. B. Sinha, S. Srivasta, *Theory of Semigroups and Applications*, Hindustan Book Agency (2017).

[92]   J. M. Steele, *Stochastic Calculus and Financial Applications*, Springer-Verlag (2001).

[93]   E. M. Stein, On the maximal ergodic theorem, *Proc. Nat. Acad. Sci.* **47**, 1894–7 (1961).

[94]   E. M. Stein, R. Shakarchi, *Fourier Analysis: An Introduction,* Princeton University Press (2003).

[95]   E. M. Stein, G. Weiss, *Introduction to Fourier Analysis on Euclidean Spaces*, Princeton University Press (1971).

[96]   Z. Suchanecki, On lambda and internal time operators, *Physica A* **187**, 249–6 (1992).

[97]   M. E. Taylor, *Pseudodifferential Operators*, Princeton University Press (1981).

[98]   N. Th. Varopoulos, Hardy-Littlewood theory for semigroups, *J. Funct. Anal.* **63**, 240–60 (1985).

[99]   D. Williams, *Probability with Martingales*, Cambridge University Press (1991).

[100] N. M. J. Woodhouse, *Geometric Quantization*, Oxford University Press (1992).

[101] K. Yosida, *Functional Analysis* (sixth edition), Springer-Verlag (1980).

[102] J. Zabczyk, *Topics in Stochastic Processes*, Scuola Normale Superiore, Pisa (2004).

[103] V. A. Zagrebnov, *Topics in the Theory of Gibbs Semigroups*, Leuven Notes in Mathematical and Theoretical Physics Vol. 10, Leuven University Press (2003).

Index

α-stable laws, 69
ω-contractive semigroup, 36

A-bound, 116
abstract Kato condition, 126
adapted process
  continuous time, 203
  discrete-time, 203
AO semigroup, 129, 156
  definition of, 11

Banach lattice, 181
Banach space
  reflexive, 88
Banach space integral, 16
Bernoulli shift, 165
boundary value problem, 38
  elliptic, 38
Brownian motion, 50–52, 73, 99, 111, 124,
    126, 130, 145, 150, 162–164, 178
  and Gauss–Weierstrass function, 51
  as martingale, 203
  definition of, 50
  rescaled, 71
  transience, 82
  with drift, 52, 63, 72
bump function, 185

C_0-group, 43, 91
  isometries, 154
C_0-semigroup
  definition of, 11
  resolvent of, 25–29
  self–adjoint, 169
Caputo derivative, 167
Cauchy distribution, 52, 71, 76

  as stable law, 69
Cauchy process, 54, 73
  rescaled, 71
Cauchy, Augustin-Louis, 166
Cayley transform, 100
Chapman–Kolmogorov equations, 130
characteristic function, 49, 57
Chernoff's product formula, 122, 127
closed graph theorem, 20
coarse graining, 164
compact operator, 102
complete positivity, 97
compound Poisson process, 72
compound Poisson semigroup, 62
conditional expectation, 200–202
  definition of, 201
conservativity, 97
contraction semigroup, 86, 117
  adjoint, 83
  definition of, 11
  for convolution semigroup, 60
  for heat equation, 50
  self-adjoint, 29, 85, 86
convolution, 186
  of functions, 47, 186
  of measures, 55–57
convolution group, 93
convolution semigroup, 46–81, 129, 140, 147,
    149
  definition of, 60
  generator, 73
  of Lévy process, 72
  on circle, 109–112
  self-adjoint, 89–91
  stable, 69
Coulombic potential, 95

Courrège theorem, 134, 136–142
  proof of, 137
cylinder sets, 124, 195

density, *see* probability density function
density matrix, *see* density operator
density operator, 112
derivation, 96
  inner, 96
differential equation
  (solution) flow, 157
  of C^k-diffeomorphisms, 158
differential equations in Banach space, 18–20
differential operator, 76
diffusion operator
  uniformly elliptic, 175
dilation, 164
Dirac, Paul, 161
Dirichlet boundary conditions, 178
Dirichlet form, 174
Dirichlet problem, 53, 54
dissipative, 117, 133
dissipativity, 29
distribution, 135–136
  of order N, 135
  support, 135
distribution tempered, 187
divergence, 159
divergence form, 38
Doeblin, Wolfgang, 205
dynamical system, 151, 160, 165
  consistency property, 154
  continuously quasi-invariant, 153, 159
  definition of, 152
  differentiably quasi-invariant, 153, 159, 160
  ergodic theory, 156
  quasi-invariant, 153

elliptic PDE, 25, 77
energy estimates, 40
entropy, 151–152
ergodic, 151
ergodic theorem, 181
ergodic theory, 164
essentially skew-adjoint, 159, 160

Feller process, 126, 131, 144, 146, 164
Feller semigroup, 126, 133, 140–142, 144,
    145, 151, 179
Feller, William, 179
Feller–Miyadera–Phillips theorem, 33, 44

Feller's pseudo-Poisson process, 131
Feynman, Richard, 123
Feynman–Kac formula, 123–126, 147
filtration
  continuous-time, 203
  discrete-time, 203
flow, 160
Fock space, 100
Fourier inversion, 188
Fourier transform, 49, 186–189
  definition of, 186
Fourier transformation, 186
Fréchet space, 187
fractional calculus, 166–167
fractional derivative, 166
Friedrichs extension, 87, 104, 175
Friedrichs–Lax–Niremberg theorem, 41
Friedrichs' theorem, 87
Friz, Peter, 175
fundamental solution
  to heat equation, 49

Gårding's inequality, 40
gamma distribution, 70
Gauss–Weierstrass function, 46, 49, 71
  as stable law, 69
generator
  as closed operator, 20–21
  definition of, 16
  of convolution semigroup, 73–76
Gibbs density operator, 113
Gibbs semigroup, 108
Gibbs state, 114
Gleason's Theorem, 114
graph norm, 20

Hamilton's equations, 160
Hamiltonian, 93, 113, 160
Hamiltonian mechanics, 160
Hardy, Godfrey Harold, 168
Hardy–Littlewood–Sobolev inequality, 168,
    173
harmonic function, 53
Hausdorff dimension, 171
heat equation, 48
heat kernel, 46, 76, 81, 178
  Gaussian bounds, 178
heat semigroup, 168
Heisenberg equation, 96
Heisenberg evolution, 113
Heisenberg picture, 95

Hellinger–Toeplitz theorem, 101
Hilbert–Schmidt operator, 104
Hilbert–Schmidt theorem, 102
Hille, Einar, 179
Hille–Yosida theorem, 36
Hille–Yosida–Ray theorem, 134, 144
hyperbolic PDE, 42

independent increments, 50, 54
infinitesimal generator
    see generator, definition of, 16
integral operator, 105
invariant measure, 149–152, 160
    ergodic, 151
    Gaussian, 150
inverse mapping theorem, 25
Itô correction, 100
Itô process, 207
Itô's formula, 51, 81, 123, 125
    proof of, 209
Itô, Kyosi, 205

Jacobian determinant, 160

K-system, 164
Kac, Mark, 123
Kato–Rellich theorem, 120
Kolmogorov consistency conditions, 195
Kologorov construction theorem
    stochastic processes, 195
Koopman operator, 155, 164
Koopman, Bernard O., 156
Koopmanism, 152, 156

Lévy continuous, 142
Lévy density, 142
Lévy kernel, 140, 142
Lévy measure, 64, 137
Lévy operator, 76, 140
Lévy process, 71–73, 130, 140, 149
    definition of, 71
"Lévy–Khintchine" form, 140
Lévy–Khintchine formula, 64–69, 75, 141
    proof of, 65
Laplace operator
    see Laplacian, 46
Laplacian, 46, 76, 174
Lax–Milgram theorem, 40
    proof of, 45
Lie, Sophus, 120
Lie–Kato–Trotter Product formula, 120–123

Lindblad equation, 114
Lindblad generator, 99
linear operator
    adjoint, 83, 88
    closable, 21
    closed, 20
        resolvent of, 25
    compact, 102, 192
    core, 22
    densely defined, 15
    dissipative, 36, 133
    domain of, 15
    essentially self-adjoint, 84, 95, 116
    extension, 15
    graph of, 20
    Hilbert–Schmidt, 104
    relatively bounded, 116
    restriction, 15
    self-adjoint, 84
        positive, 85
    symmetric, 84
    trace class, 105
    unbounded, 15
Liouville's theorem, 160
Lipschitz condition
    global, 157, 158
    local, 158
Littlewood, John Edensor, 168
Lorentz equations, 158
Lumer–Phillips theorem, 36, 86, 117

manifold, 158
    orientable, 160
    Riemannian, 171, 174
Markov kernel, 129, 130
Markov operator, 156
Markov process, 130, 131, 147, 178
Markov property, 130
Markov semigroup, 130, 164, 165
    invariant measure, 149
martingale, 145, 203–204
    centred, 145
    continuous-time, 203
    discrete-time, 203
martingale problem, 144–146
measure
    inner regular, 212
    outer regular, 212
    regular, 212
Mehler semigroup, 150
Mellin transform, 169

Mercer's theorem, 107, 111
Miyadera, Isao, 179
monoid, 14
  topological, 14

Nash inequality, 175
Newton's laws, 161
Newtonian potential, 82

observables, 95
one-parameter unitary group, 93, 99, 159
operator-valued stochastic differential
    equation, 99
Ornstein–Uhlenbeck process, 150

parabolic PDE, 9, 38–43, 104
Parseval identity, 187
  proof of, 188
path space, 124
Perron–Frobenius operator, 154–156, 159
phase multiplication operator, 186
phase space, 160
Phillips, Ralph, 179
Plancherel theorem, 187
  proof of, 188
Planck's constant, 93
Poisson bracket, 160
Poisson kernel, 53
Poisson summation formula, 112
portmanteau theorem, 57
positive maximum principle, 133–144
  definition of, 133
positive self-adjoint operator, 85
Post–Widder inversion formula, 127
potential, 93, 123, 126
Povzner–Nelson theorem, 160
Prigogine, Ilya, 164
probability density function
    as Radon–Nikodym derivative, 200
probability kernel, 129
probability space
  definition of, 194
pseudo-differential operator, 64, 74, 76–78,
    80, 141, 144, 146
  definition of, 77
  symbol, 77

quantum dynamical semigroups, 97–100, 164
quantum Liouville equation, 113, 114
quantum mechanics, 93, 112, 116, 161
quantum partition function, 114

quantum statistical mechanics, 114
quasi-invariant, 152

Radon–Nikodym derivative, 56, 81, 150, 153,
    154, 160
  definition of, 200
Radon–Nikodym Theorem
  proof of, 197
random variable
  definition of, 194
  distribution of, 194
  law of, 194
  integrable, 200
relatively bounded, 116
relativistic Schrödinger operator, 82, 101
Rellich–Kondrachov theorem, 104, 192
resolvent
  compact, 115
resolvent density, 82
resolvent set
  definition of, 25
Riemann, Bernhard, 167
Riemann–Liouville fractional derivative, 167
Riemann–Liouville integral, 167
Riesz lemma, 135, 213
Riesz potential
  generalised, 172
Riesz potential operator, 168, 169
  generalised, 173
Riesz representation theorem, 39, 40, 45, 213
  statement of, 197
Riesz–Thorin interpolation theorem, 169, 178
rotationally invariant measure, 70

Schrödinger equation, 93, 94, 123, 161
Schrödinger evolution, 113
Schrödinger picture, 95
Schwartz space, 73, 187
semidynamical system, 13, 151–158, 180
  invariant measure, 152
    ergodic, 156
  definition of, 152
semiflow, 158
semigroup
  analytic, 119–120
  bounded, 120
  compact, 102–104
  conservative, 131, 147
  dilation, 163
  Feller, 179, 181
  Markov, 130, 181

norm-continuous, 23–25
positive, 131, 181
self-adjoint, 162
symmetric diffusion, 104
trace class, 104, 108
ultracontractive, 170, 178
Varopoulos, 166
separating
subalgebra, 185
Sobolev embedding theorem, 191
Sobolev inequality, 174
Sobolev space, 39, 81, 190–193
anisotropic, 80
of order m
definition of, 190
Sobolev, Sergei, 168
spectral inclusion theorem, 180
spectral mapping theorem, 180
spectral radius, 127
spectral theorem, 84, 99
spectrum
definition of, 25
stable semigroup, 69–71, 76
state, 95
mixed, 112
pure, 112
Stein's maximal ergodicity lemma, 171
stochastic differential equation, 131, 145, 178
stochastic integral
definition of, 205
stochastic process
continuous time, 194
definition of, 194
discrete time, 194
finite-dimensional distributions, 194
Stone's theorem, 94
Stone–Weierstrass theorem, 185
sub-Feller semigroup, 147
subordinator, 69
symbol, 77, 91

Sz-Nagy, Bela, 164

tempered distributions, 187
theta function, 112
topological semigroup, 14
trace class operator, 105
transition density, 150
transition probability, 130, 149
transition semigroup, 130, 149, 178
translation group, 93
translation operator, 186
translation semigroup, 13, 30, 63, 76

uniform boundedness principle, 11
uniform exponential stability, 180
uniform stability, 180
uniformly elliptic, 9
unitary operator-valued stochastic process, 99
Uruysohn's lemma, 185

vanish nowhere
subalgebra, 185
Varopoulos semigroup, 166–178
Varopoulos's theorem, 173
Varopoulos, Nicholas, 169
vector field, 158
complete, 158, 160
divergence, 159
volume form, 160
volume measure, 160

weak convergence
of measures, 57–59
weak solution
to elliptic BVP, 39
Wiener measure, 123, 124

Yosida approximants, 31–33, 44
Yosida, Kosaku, 179